Studies in Big Data

Volume 19

About this Series

The series "Studies in Big Data" (SBD) publishes new developments and advances in the various areas of Big Data- quickly and with a high quality. The intent is to cover the theory, research, development, and applications of Big Data, as embedded in the fields of engineering, computer science, physics, economics and life sciences. The books of the series refer to the analysis and understanding of large, complex, and/or distributed data sets generated from recent digital sources coming from sensors or other physical instruments as well as simulations, crowd sourcing, social networks or other internet transactions, such as emails or video click streams and other. The series contains monographs, lecture notes and edited volumes in Big Data spanning the areas of computational intelligence incl. neural networks, evolutionary computation, soft computing, fuzzy systems, as well as artificial intelligence, data mining, modern statistics and Operations research, as well as self-organizing systems. Of particular value to both the contributors and the readership are the short publication timeframe and the world-wide distribution, which enable both wide and rapid dissemination of research output.

More information about this series at http://www.springer.com/series/11970

Dominik Ryżko · Piotr Gawrysiak
Marzena Kryszkiewicz · Henryk Rybiński
Editors

Machine Intelligence and Big Data in Industry

Springer

Editors
Dominik Ryżko
Institute of Computer Science
Warsaw University of Technology
Warsaw
Poland

Marzena Kryszkiewicz
Institute of Computer Science
Warsaw University of Technology
Warsaw
Poland

Piotr Gawrysiak
Institute of Computer Science
Warsaw University of Technology
Warsaw
Poland

Henryk Rybiński
Institute of Computer Science
Warsaw University of Technology
Warsaw
Poland

ISSN 2197-6503 ISSN 2197-6511 (electronic)
Studies in Big Data
ISBN 978-3-319-30314-7 ISBN 978-3-319-30315-4 (eBook)
DOI 10.1007/978-3-319-30315-4

Library of Congress Control Number: 2016932358

Printed on acid-free paper

This Springer imprint is published by Springer Nature
The registered company is Springer International Publishing AG Switzerland

Preface

This book presents valuable contributions devoted to practical and, in many cases, industrial applications of Machine Intelligence and Big Data in various branches of the industry. All the contributions are extended versions of presentations delivered at the Industrial Session of the 6th International Conference on Pattern Recognition and Machine Intelligence (PREMI 2015) in Warsaw, Poland, which passed through a rigorous reviewing process. Each paper was reviewed by at least two referees.

Part I is focused on practical applications of text processing. This part demonstrates the usefulness of text mining approaches in solving practical problems. In particular, Sobkowicz addressed the problem of automatic sentiment analysis for Polish language. Kowalski presented a process of translating legal English and Polish phrases, being a part of a bilingual university repository. Roziewski et al. studied creation of *n*-gram collection from a large-scale corpus of Polish Internet based on Common Crawl Corpus. Kozłowski applied clustering of documents containing Polish national qualifications frameworks in order to analyze study fields. Various approaches to the semantic textual similarity are compared in the work by Kazuła and Kozłowski.

Part II is devoted to data mining. El-Baz et al. proposed a solution of the problem of identification of diabetes disease by means of committees of neural network-based classifiers. Sharif et al. proposed enzyme function classification based on Borda count ranking aggregation method. The problem of mining of frequent action rules is addressed by Dardzinska and Romaniuk.

Text and multimedia processing is the subject of Part III. Protaziuk et al. proposed an automatic machine translation method for translating multi-word labels from lexical layers of domain ontologies. In the area of automated speech recognition, Madhavi et al. addressed vocal tract length normalization using different warping functions for template matching. A comparative study on music genre classification algorithms was presented by Stokowiec.

Issues of software platforms are studied in Part IV. Blachnik and Kordos described a RapidMiner Library for information selection and data compression.

Wróblewska et al. showed how to cluster offers in an e-commerce marketplace in order to improve performance of recommendations and other services. An application of machine learning algorithms to Bitcoin automated trading is described by Żbikowski.

Part V combines papers on complex systems, the Internet of Things, and agent systems. Kopczynski et al. presented a design for hardware cuts generating module for Field Programmable Gate Arrays (FPGAs). A big data solution for smart grids and smart meters was presented by Konopko. Weclawski and Jankowski presented an intelligent system of limited resource allocation for large-scale agent systems. Yadav et al. studied the problem of finding logical patterns in multi-sensor data from the industrial Internet.

We thank all the authors for their contributions to the book and we express our appreciation for the work of the reviewers. We express our gratitude to the industrial partners: mBank, Allegro, and Samsung for their financial support to the PReMI 2015 conference and to this publication.

November 2015

Dominik Ryżko
Piotr Gawrysiak
Marzena Kryszkiewicz
Henryk Rybiński

Contents

Part I
Text Processing

Automatic Sentiment Analysis in Polish Language

Antoni Sobkowicz

Abstract We introduce fully automated process for sentiment analysis in short texts in Polish language. Process consists of (a) generation of emotion lexicon using Twitter annotated messages (b) building sentiment data set using annotated messages and the generated lexicon, (c) training NEAT genetic algorithm using previously prepared data set and (d) the final evaluation using 10 fold cross validation. We show that this method provides good results and can be used to simplify sentiment analysis processes for Polish language content.

Keywords Sentiment analysis · NLP · Machine learning

1 Introduction

Sentiment analysis is a process of evaluating sentiment (or polarity) of a text—for example, if a review was positive or negative one. Automatic detection of sentiment has variety of uses—from determining response to political statements to monitoring customer satisfaction based on social media. While good-quality, similar systems exist for English language, tools for Polish language are not very good, mostly because of poor sentiment lexicon base. Our approach automates (a) generation of huge sentiment lexicons and (b) analysis of text using machine learning algorithm with features based on generated lexicons and grammar. Automated lexicon generation allows not only to sentiment-tag commonly used words, but also to tag sentiment of rarely used ones, and specific dialects. Generated lexicon is used to analyze texts taking into account negation, adjectives, adverbs, and create Sentiment Vector for text. Sentiment Vectors are used both to train analysis NEAT neural network algorithm, and used afterwards to determine text sentiment. Algorithm evaluation is performed using 10-fold cross validation using part of data exploited during lexicon

A. Sobkowicz (✉)
Natural Language Processing Laboratory, National Information Processing Institute,
Niepodleglosci 188 B Avenue, 00-608 Warsaw, Poland
e-mail: antoni.sobkowicz@opi.org.pl
URL: http://www.opi.org.pl/en/Natural-Language-Processing-Lab.html

D. Ryżko et al. (eds.), *Machine Intelligence and Big Data in Industry*,
Studies in Big Data 19, DOI 10.1007/978-3-319-30315-4_1

generation (around 20 % of all data). We obtained accuracy of 0.85. We also tested algorithm, which was trained on partial data, on full dataset to see how algorithm works on unknown data and we obtained accuracy of 0.80. We also compared these results to Naive Bayes machine learning based algorithm (accuracy 0.72 and 0.56 respectively).

2 Related Work

Sentiment analysis for English language algorithms is a heavily explored topic, with many already existing, commercially available tools, like Sentistrength [1]. Analysis of short messages is the topic of Conference on Semantic Evaluation Exercises (SemEval), where different approaches are tested on prepared dataset. On SemEval-2013 [2], best performing algorithms were based on machine learning and variety of features extracted from text. Most of presented systems benefited from existing Sentiment Lexicons. Automatic sentiment lexicon generation from short messages (Twitter) was explored by Kiritchenko et al. [3], with good results. Sentiment lexicons for Polish language exists, but are not top quality and are missing large amount of data. Because of that, sentiment analysis using those lexicons do not provide acceptable results. Sentiment dictionaries made for specific communities work well [4], but their usability is limited to their origin communities. Approaches to automatic lexicon generation and/or sentiment analysis for polish language [5, 6] are based on product reviews from e-commerce websites like ceneo.pl and Internet shops.

3 Approach

3.1 Emotion Lexicon Generation

To create usable and large corpus we decided to use automated process similar to one described in [3]—we used automated tools to gather Polish Twitter data for positive and negative emotions using build in Twitter filtering mechanism. To retrieve positive sentiment input data, we filtered out Twitter stream with "lang:pl :)" which, according to Twitter, detects text with positive emotions (using mostly emoticons as authors observed). To retrieve negative sentiment data, we filtered out stream using "lang:pl :(wkurwiony spierdalaj wypierdalaj jebany lewaki katotalib", which detects both Twitter annotated negative emotions (":(") and popular polish swear words and two non-swear negative terms ("lewaki"—eng. leftists, negative term used for describing ultraliberals and "katotalib"—eng. catho-talibans, negative term used for describing ultracatholics). As negative input data was a lot smaller than positive input data, we supplemented it by gathering content from site wkurwiamnie.pl, which contains texts about things that make people angry and infuriated.

Web scrapping software was written in Python using Tweepy library for twitter part, and Scrapy for wkurwiamnie.

During two-week crawling period we gathered around 140 000 positive texts and 30 000 negative texts. For each text with emotion e, where $e = 1$ for positive emotion and $e = -1$ for negative emotion, we tokenized text. Each token was lemmatized using Morfologik, and if lemma was a correct polish word, it was added to collection W, which held all found lemmas. Lemma that has been found was also added to either E_1 or $E_{=1}$ collection, depending on text emotion. We also added bigrams and non-connected bigrams of lemmas (two lemmas separated by another).

After all texts were analyzed, we calculated sentiment score s for each lemma t in collection W. Sentiment score was calculated as shown below (after [3]):

$$s(t) = PMI(t, positive) - PMI(t, negative) \tag{1}$$

PMI stands for pointwise mutual information

$$PMI(t, positive) = log_2 \left(\frac{freq(t, E_1) * count(W)}{freq(t, W) * count(E_1)} \right) \tag{2}$$

where $freq(t, E_1)$ is the number of times token t occurs in collection E_1 and $count(W)$ is number of different tokens t in collection W. $freq(t, W)$, $freq(t, E_{-1})$ are described in similar way, as is $PMI(t, negative)$. Equation (1) can be simplified to:

$$s(t) = log_2 \left(\frac{freq(t, E_1) * count(E_{-1})}{freq(t, E_{-1}) * count(E_1)} \right) \tag{3}$$

Using described procedure, we generated dictionary containing 370 000 phrases—single lemmas, bigrams and non-connected bigrams.

3.2 Sentiment Analysis Algorithm

All texts used for NEAT genetic algorithm training and validation were analyzed using custom algorithm. Algorithm worked as follows:

1. Data preparation
2. Split text into sentences, using .?! as delimiters
3. Tokenize sentences
4. Lemmatize word, and get its lemma and tags
5. Check for special lemmas
6. Calculate local lemma sentiment value s_f
7. Increment global text sentiment value t_f and calculate Sentiment Vector

Each step of algorithm is described in detail below.

Data preparation. Before text was processed, it had been divided into sentences using ".", "?" and "!" as delimiters. This separation allowed us to make special lemmas (like negation lemma) affect only one sentence. Then, each sentence was divided into words wd using " " (space) as delimiter, and each word wd was lemmatized using lemmatizer based on Morfologik. Resulting lemma lm was saved and lemma tags tg_{lm} for lemma lm were retrieved using Morfologik based tool.

Special lemmas. Next, we detected if lemma lm is some kind of special meaning using its tags, tg_{lm}, such as negation or gradation. These kind of lemmas, in addition to carrying sentiment value, change meaning of either whole sentence or next lemma and bigrams. Below is list of all detected lemmas and modifiers introduced by them:

Negation lemma changes following sentiment values to negative (from positive) or less negative. We set negation flag f_n to 1. There is one negation lemma, "nie" (pol. no).

Tags "verb" mean that lemma is a verb. We set power modifier m to $m = 1.5$

Tags "adj" and "com" mean that lemma is comparative adjective. We set power modifier m to $m = 2$. We also set sentiment push flag f_p to 1, as adjectives modify sentiment of lemmas following them, and store lemma sentiment $s_s = s(lm)$. Example—najlepszy (pol. best).

Tags "adj" and "sup" mean that lemma is superlative adjective. e set power modifier m to $m = 3$. We also set sentiment push flag f_p to 1, as adjectives modify sentiment of lemmas following them, and store lemma sentiment $s_s = s(lm)$. Example—lepszy (pol. better).

Tags "adj" mean that lemma is an adjective. e set power modifier m to $m = 1.5$. We also set sentiment push flag f_p to 1, as adjectives modify sentiment of lemmas following them, and store lemma sentiment $s_s = s(lm)$. Example—dobry (pol. good).

Gradation lemma changes gradation of adjective or adverb following it. We set superlative flag f_s to 1.

Calculate lemma sentiment value. First, we calculate base lemma lm sentiment s. All lemmas lm encountered when flag $f_n = 1$ will have their sentiment s modified: $s = s - s(lm)$ if $s(lm) > 0$ and $s = s(lm) * 0.6$ if $s(lm) < 0$, as negating already negative lemmas do not change them into positive, only changes them into less negative. If flag $f_n = 0$, lemma sentiment s is equal to lexicon sentiment value, $s = s(lm)$.

Next, we apply push sentiment flag f_p if applicable, and calculate final modifier m_f. If push sentiment flag $f_p = 1$, modifier m_f becomes $m_f = m * 1.2$ if $s_s * s > 0$ and $m_f = m * 0.6$ if $s_s * s < 0$. If push sentiment flag $f_p = 0$, modifier $m_f = m$. We also add $0.1 * s_s$ to lemma sentiment s if $f_p = 1$.

We also check if previous token and current token don't form a bigram bi that has sentiment value stored in lexicon. If yes, we get base bigram sentiment $b = s(bi)$ calculate bigram sentiment b_c:$b_c = b * m * -0.4$ if negate flag $f_n = 1$ or $b_c = b *$

$m * 0.24$ if negate flag $f_n = 0$. We do the same for current token and token 2 places before it, getting non-connected bigram sentiment value b_{nc}.

Having calculated both final modifier m_f and lemma sentiment s and bigram sentiments b_c and b_{nc}, we can calculate final word sentiment s_f:

$$s_f = log_2 \left(\frac{freq(t, E_1) * count(W)}{freq(t, W) * count(E_1)} \right) \quad (4)$$

Calculate text sentiment value and sentiment vector. To get final sentiment value for text t_f, we add lemma sentiment value s_f calculated for each lemma to t_f. We also construct Sentiment Vector sv, which consists of:

- Final sentiment t_f
- All tokens c
- Number of tokens found c_t
- Number of tokens with positive final sentiment c_p
- Number of tokens with negative final sentiment c_n
- Sentence count c_s
- Highest positive token final sentiment value t_p
- Lowest negative token final sentiment value t_n

We store Sentiment Vector sv in collection SV.

3.3 Sentiment Analysis Algorithm

Sentiment analysis algorithm is based on NEAT neural network. NEAT (Neuroevolution of augmenting topologies) is a genetic algorithm for generating evolving neural networks. It was created by Stanley O. Kenneth and is described in [7].

This part of process was implemented using Encog library for Java language, developed by Jeff Heaton.

Training Neural Network. To train neural network, we selected training dataset—6 files containing 14 053 negative texts total and 6 files containing 16 048 positive texts. Difference in positive and negative text count was mostly due to Having obtained Sentiment Vectors for all text in dataset, we saved them to CSV file for use with training algorithm, along with information about original text sentiment.

We loaded data using Encog library tools, and trained NEAT neural network using 90 % of training dataset, using 10 % to validate training, using 10-fold cross validation for model tweaking. From 10 folds, optimization algorithm build into Encog chose model performing best, and saved model for future use.

4 Analysis

To check sentiment analysis process, we performed 3 tests comparing our NEAT neural network based algorithm to algorithms using FeedForward networks, Support Vector Machines, Deep Belief Networks and Naive Bayes (as a baseline):

- 10-fold cross validation using training dataset (30 101 texts total, 14 053 negative and 16 048 positive)
- Validation using full dataset, with training dataset using for training (around 170 000 texts total for validation, 27 090 used for training)
- 10-fold cross validation using extended training dataset (around 170 000 texts total, 130 000 positive and 40 000 negative).

10-Fold Cross Validation We performed 10-fold cross validation using training dataset. For validating NEAT neural networks, FeedForward networks and Support Vector machines, we used mechanism build into Encog library (also used in network training procedure). For Naive Bayes, we used library developed by Philipp Nolte with our extensions. For Deep Belief Networks, we used ml4j library. Results for 10-fold cross validation as shown in Table 1.

Validation Using Full Dataset We performed validation using full dataset and models trained on training dataset, to see how different models performed with new data (Table 2).

10-Fold Cross Validation Using Extended Training Dataset We performed 10-fold cross validation using extended training dataset, consisting of all data gathered. This allowed us to see how model accuracy changed with larger training data, as neural networks, are prone to overfitting, and Naive Bayes based models are more effective larger the training set get (Fig. 1). For validating NEAT neural networks we

Table 1 10-fold cross validation for training dataset

	Method				
Fold	NEAT	Naive Bayes	FeedForward	SVM	DBN
1	0.828	0.723	–	0.810	–
2	0.864	0.730	–	0.809	–
3	0.851	0.735	–	0.787	–
4	0.873	0.719	–	0.802	–
5	0.856	0.726	–	0.810	–
6	0.820	0.718	–	0.808	–
7	0.828	0.730	–	0.808	–
8	0.870	0.729	–	0.798	–
9	0.852	0.721	–	0.815	–
10	0.860	0.728	–	0.803	–
Avg	0.85	0.72	0.50	0.81	0.599

Table 2 Full set validation

Method	Accuracy
NEAT	0.80
Naive Bayes	0.56
FeedForward	0.78
SVM	0.77

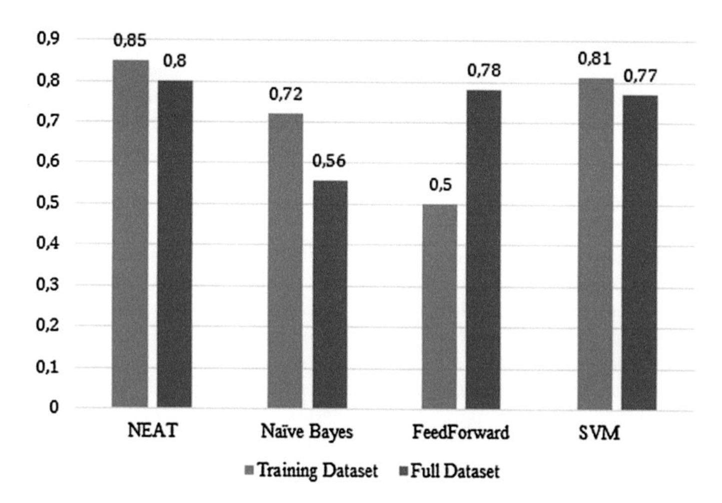

Fig. 1 10-cross fold results for training dataset validation only and validation using full dataset. All methods (except FeedForward neural network based) show degraded performance

Table 3 10-fold cross validation for extended training dataset

Method	Accuracy	Accuracy on full dataset
NEAT	0.83	0.82
Naive Bayes	0.72	0.74

used mechanism build into Encog library (also used in network training procedure). For Naive Bayes, we used library developed by Philipp Nolte, with our extensions. Results for 10-fold cross validation as shown in Table 3.

5 Conclusion

With accuracy of 0.82, using 10-cross fold when whole dataset was used for validation and accuracy of 0.80 when working on full dataset, with learning set being only 15 % of full dataset, we can safely say that proposed algorithm works very well despite having limited input dataset. Ability to automate training set generation makes this algorithm easy to use. Future improvements to algorithms are under consideration—features such as emoticon detection and introducing agent-based emotion tracking system for keeping previous text emotions—may make algorithm better suited for working on smaller communities, where agent interactions are important.

References

1. Thelwall, M., Buckley, K., Paltoglou, G., Cai, D., Kappas, A.: Sentiment strength detection in short informal text. J. Am. Soc. Inf. Sci. Technol. **61**, 2544–2558 (2010)
2. Wilson, T., Kozareva, Z., Nakov, P., Rosenthal, S., Stoyanov, V., Ritter, A.: SemEval-2013 task 2: sentiment analysis in Twitter. In: Proceedings of the International Workshop on Semantic Evaluation (2013)
3. Kiritchenko, S., Zhu, X., Mohammad, S.: Sentiment analysis of short informal texts. J. Artif. Intell. Res. **50**, 723–762 (2014)
4. Sobkowicz, P., Sobkowicz, A.: Two-year study of emotion and communication patterns in a highly polarized political discussion forum. Soc. Sci. Comput. Rev. **30**, 448–469 (2012)
5. Haniewicz, K., Rutkowski, W., Adamczyk, M., Kaczmarek, M.: Towards the Lexicon-Based Sentiment Analysis of Polish Texts: Polarity Lexicon. Lecture Notes in Computer Science, vol. 8083. Springer, Berlin (2013)
6. Buczyski, A., Wawer, A.: Automated classification of product review sentiments in polish. Intelligent Information Systems, pp. 213–217 (2008)
7. Miikkulainen, R., Kenneth, S.: A taxonomy for artificial embryogeny. Artif. Life **9**, 93–130 (2003)

Learning Curve with Machine Translation Based on Parallel, Bilingual Corpora

Maciej Kowalski

Abstract Machine Translation is a branch of computer science that automatically handles translation of a text from a source language to a target language. This article summarizes the experience gained during *UKSW* project, part of which deals with translation of legal phrases between English and Polish. The article describes consecutive steps of the project, i.e. collecting data and creating parallel, bilingual corpora, checking open source ready-made solutions and the novel, effective *SMT* solution that has been proposed. The final chapter summarizes the solution, together with the results based on *BLEU* metrics.

Keywords Machine translation · Bilingual corpora · Parallel corpora

1 Background

Machine translation is a part of computer linguistics, which is focused on using automatic methods and computer algorithms in order to translate a text from one language to another. It is nowadays used widely in various areas both professional, like translating technical documents, instruction, reports [1, 2], as well as everyday, such as communicating with other people by using different than mother language,[1] translating pages at run-time, teaching and learning.[2]

Considering both of those applications, many of the current state-of-the-art systems require to be reviewed by human translator who corrects the translation to be

[1]http://translate.google.pl.
[2]http://duolingo.com.

M. Kowalski (✉)
National Information Processing Institute, al. Niepodleglosci. 188B,
00-608 Warsaw, Poland
e-mail: mkowalski@opi.org.pl
URL: http://www.opi.org.pl

D. Ryżko et al. (eds.), *Machine Intelligence and Big Data in Industry*,
Studies in Big Data 19, DOI 10.1007/978-3-319-30315-4_2

more humane. As translation methods and existing solutions are getting better, there is still a need of human interaction and overseeing.

Machine Translation as it is today, may implement one or more of the tree following methods:

- rule-based, also known as knowledge-based translation, which depends on information about two languages having their place in the process. To be able to use this method, one must first analyze both of the languages to find grammar, semantic, morphological or syntactic regularities. After this step, two dictionaries are build for words and rules. Based on the information found, the translation process tries to link the input sentence structure with the output sentence, filling the translation with a data coming from the dictionary.
- example-based (statistical, case-based), which depends on bilingual corpora, to generate statistical model of a language. Given phrase is translated according to probability distribution that an output phrase the translation of it. One of the simplest implementation applies Bayes Theorem.
- hybrid-based, which uses multiple approaches within one machine translation system. The key reason why hybrid approach is being used, is that it has better accuracy than separate approaches.

2 Problem Description

One of the key features of *UKSW* project, was to aid law-related people with translating legal phrases from Polish to English and vice versa. The *UKSW* project consists of several parts, which work together in following way (as an example):

- a person is searching for key phrases, which occur in official court rulings,
- court ruling is described by law acts that are incorporated into it. As a result, user is given not only the text of the ruling, but also all the references to the law acts.
- user can access particular law act and find related court rulings to it.

As it was learned during the *UKSW* project, lawyers or solicitors often make examples or try to find reasoning for their cases in court. In each step of this process, machine translation is required to be working and be available for users, so foreign court ruling and acts may be accessed. That way users are able to find similar cases in other countries.

2.1 Limitations

The *UKSW* project has been limited by two important factors, which are memory size and disk space that the developer/production systems were installed on. Typical machine contained Intel(R) Core(TM) i7-3770 CPU @ 3.40 GHz with 16 GB of

RAM and 300 GB of disk space. Using already proven systems were far more resource-consuming:

- Joshua[3]—RAM consumption depends on amount of parallel data, which the system is trained on. Initial configuration suggest more than 32 GB of RAM.
- Moses[4]—RAM consumption is estimated to be 8 GB for 200,000 items of training data and 100 GB of disk space. Authors suggest that amount of disk space should be 100 times the amount of learning data.

In both cases bottlenecks are RAM or disk space usages, which both excludes those frameworks from being the solution to described problem.

2.2 Data

The translation system described in this article uses parallel bilingual data (Polish and English). The total amount of Polish-English data is approximately 42,000,000 items of words, phrases, sentences and whole documents (different granularity). Data were incorporated from following sources:

- European law acts database—*EUPARL*,[5]
- *TED* talks transcriptions,[6]
- corpus data available at Institute of English Studies at Lodz—*ACADEMIA*, *OSW*, *CORDIS*, *JRC*, *RAPID*, *ESO*, [7]
- dictionaries—*DICT*,[8,9]
- paralell corpora: movies subtitles, medical documentation, manuals, user interfaces of OpenOffice software *DGT*, *EAC*, *ECDC*, *OPUS*, *LIT*,[10]
- eurlex law acts, court orders—*EURLEX*[11]
- European Court of Justice—*CURIA*[12]

For given table (Table 1), $LEN_{min}(EN)$ and $LEN_{max}(EN)$ are the shortest and longest lengths of text blocks found in English part for *source*, $LEN_{min}(PL)$ and $LEN_{max}(PL)$ are the shortest and longest lengths of blocks found in Polish part for *source*. *Count* is the total number of pairs found in *source*.

[3]http://joshua-decoder.org/6.0/pipeline.html.

[4]https://github.com/jladcr/Moses-for-Mere-Mortals.

[5]http://www.statmt.org/europarl/.

[6]http://www.ted.com/talks/browse.

[7]http://pelcra.pl/new/.

[8]http://www.slowniki.org.pl/.

[9]http://www.diki.pl/.

[10]http://opus.lingfil.uu.se/index.php.

[11]http://eur-lex.europa.eu/.

[12]http://curia.europa.eu/.

Table 1 Parallel bilingual data statistics

Source	$LEN_{min}(EN)$	$LEN_{max}(EN)$	$LEN_{min}(PL)$	$LEN_{max}(PL)$	Count
EURLEX	5	1,009,682	0	927,844	531,938
CURIA	0	4595	0	4669	495,924
DGT	31	91,866	30	86,364	2,652,699
EAC	31	1147	31	1123	1699
ECDC	0	1437	31	1512	1970
EUPARL	3	1737	0	1938	629,322
TED	9	98	6	120	3607
CORDIS	2	701	1	934	467,763
RAPID	5	1617	2	1632	371,712
OPUS	2	40,673	5	6871	35,932,031
DICT	7	62	2	201	217,410
ACADEMIA	6	1634	1	1477	17,857
ESO	6	3622	4	3773	4230
JRC	6	4324	0	4448	1,387,708
LIT	6	2210	3	2482	35,434
OSW	6	3718	0	3360	28,718
				Total	42,780,022

Table 2 Size of corpus by granularity

Granularity type	Count
Words	1,024,374
Phrases	14,419,839
Sentences	20,944,836
Documents	6,243,735

For given table (Table 2), words are strings with no white characters included, phrases are strings which have 2–5 white characters included, sentences are strings which have 6–20 white characters included and documents are all strings which contain more than 20 white characters.

All the data was stored using search and indexing engine—*Lucene*.[13] There was no language-specific analyzer used.

Example 1 Structure of *Lucene* document used to store bilingual data record

```
Document {
    <pl:75 OHIM nie zgadza sie z argumentami skarzacej.>
    <en:75 OHIM disputes the applicants arguments.>
}
```

[13]https://lucene.apache.org.

```
Document {
    <pl:77 Nalezy zgodzic sie z ta analiza.>
    <en:77 That analysis must be upheld.>
}
```

3 Process

3.1 Data Alignment

Based on the fact that the data contained entries, which size ranged from few letters up to 1 MB of text, an alignment process should be performed. Standard aligner tools like GIZA++,[14] Berkeley,[15] Hunalign[16] are time and resource consuming processes, therefore an novel alignment method has been introduced that outperforms them.

The indexes between, which result phrase may be located are computed as follows:

$$index_{min} = POS_{SRC}(word)\frac{LEN(block_{SRC})}{LEN(block_{DST})} - \delta \tag{1}$$

$$index_{max} = POS_{SRC}(word)\frac{LEN(block_{SRC})}{LEN(block_{DST})} + \delta \tag{2}$$

For the Eqs. (1) and (2), $POS_{SRC}(word)$ is the position of input phrase inside *source* text block, $LEN(block_{SRC})$ is the length of source text block, $LEN(block_{DST})$ is the length of destination text block, δ is a fixed shift which is by default 2 words.

In the following example, a translation for word "poniewaz" will be searched.

Example Alignment based on the following data block:

```
pl:37. Komisja jest zdania, ze art. 9 ust. 2 lit. dekretu
krolewskiego 14702007 jest sprzeczny z art. 42 ust. 3
rozporzadzenia nr 17822003, poniewaz narusza on zasade
rownosci traktowania.

en:37. The Commission considers that Article 92b of Royal
Decree 14702007 is not compatible with Article 423 of
Regulation No 17822003 because it is inconsistent with
the principle of equal treatment.
```

The word "poniewaz" starts at the index = 135 inside *Polish* part of text block. By using presented formulas (1) and (2), the most probable indexes, between which the translation can be found, are (an assumption is made that $\delta = 2$ words):

---

[14]http://www.statmt.org/moses/?n=FactoredTraining.RunGIZA.

[15]https://code.google.com/p/berkeleyaligner/.

[16]https://github.com/danielvarga/hunalign.

$$index_{min} = 135\frac{182}{194} - 2_{words} = 106 \quad (3)$$

$$index_{max} = 135\frac{182}{194} + 2_{words} = 139 \quad (4)$$

The most probable space of finding the correct translation is between indexes 106 and 139, which is "regulation No. 17822003 because it is".

3.2 Translation

General translation method contains following steps:

1. An input phrase is being split into n-grams of the predefined maximum size as follows:

```
Input phrase: A B C D
N-Gram max size: 3
N-Grams: A, B, C, D, A B, B C, C D, A B C, B C D
```

2. Each generated source n-gram is taken as a query to *Lucene* index. If a search has PL-EN direction, the phrase is being searched in PL part of index, otherwise, the EN part is taken into consideration.
3. Each corresponding text block is narrowed down using data alignment method described above.
4. Each narrowed text block found using *Lucene* index is processed by the tokenizer as it is described in point 1. The resulting n-grams are stored in a sorted list, which relates to source n-grams. The more frequent the resulting n-gram is, the higher in the list it is placed.
5. Translation begins from the longest n-grams. The source phrase is, step-by-step being replaced by most probable (the most appearing in results) n-grams.

Example 1:

```
PL: karty do gry

karty -> [cards(114), card(16), you(15), and(10)]
do -> [do(50), to(38), you(22), up(16)]
gry -> [game(79), over(51), games(38), is(14), playing(13)]
karty do -> [playing cards(142), you know them playing cards(140),
            cards and(129), cards of(129)]
do gry -> [back in(38), to play(30), you back in(24), to you(22)]
karty do gry -> [playing cards(166), you know them playing
cards(140), are playing cards it is(132), playing cards and(131)]

EN: playing cards
```

The number given next to the resulting n-gram is the number of coexisting occurrences of n-gram and source n-gram found, using *Lucene* and it is treated as translation score. Given translation result is then cross-checked with more granule translations (appearing higher in list). In following example, "playing cards" is split into two words: "playing" and "cards". If those two words are found in the earlier results, the final score is increased by scores of finer phrases, so in this case final score is $166 + 13 + 114 = 293$ and is the most probable translation. In this example, there exists a phrase, which is translated 1:1 with given input phrase.

Example 2:

```
PL: podejrzany o wlamanie

podejrzany -> suspect(114), is(21), suspicious(14), he(12)
o -> oh(44), it(27), is(19), yeah(15)
wlamanie -> and(50), breaking(46), entering(42), burglary(35)
podejrzany o -> suspect in(146), prime suspect in(127), robbery
suspect(127), suspect on two(122)
o wamanie -> and breaking and entering(109), breaking and
entering(109), breaking and     entering of(104), of breaking
into our(69)
podejrzany o wlamanie -> []

EN: suspect in and breaking and entering
```

In this example there is no 1:1 translation proposition, so the translation must be composed from shorter n-grams. The next n-gram to be analyzed is "and breaking and entering", which is decomposed into words "and", "breaking", "entering". For source n-gram "o wlamianie", this is the most probable translation with score: $109 + 46 + 42 = 192$, which is used as the translation of that part. After that, only "podejrzany" should be translated and the most probable n-gram for this input is suspect with score $= 114$. Overall translation is then composed into one, to be "suspect in and breaking and entering".

4 Variations and Test Results

BLEU is a metric, used for evaluation of quality of machine translated texts in comparison to reference translations. The closer to a reference the machine translation is, the higher the score it has. As an output, metric gives a real number between 0 and 1, for 1 being exact match with given reference and 0 being completely different [3].

The *BLEU* metric is first of its kind, which managed to project the correctness the translated text onto the real number between 0 and 1. Due to the simplicity of implementation and speed of operation, it is still used to measure the quality of machine translated texts. Presented method was assessed by using *BLEU*.

Table 3 Translation test cases (including common and legal phrases)

English common phrase	Polish common phrase
I want to press the point	chce odkreslic, ze
Willingly	chetnie
What I meant is	chodzi mi o to, ze
At times	chwilami
To the moment	co do sekundy
English legal phrase	Polish legal phrase
Criminal action	proces karny
Criminal proceedings	postepowanie karne
Death tax	podatek od spadku
Debt collector	windykator
Default judgment	wyrok zaoczny

Table 4 BLEU score for common phrases

	ORIGINAL	FIXED	MORFOLOGIK	EXTENDED
COMMON	0.48	0.48	0.22	**0.52**
LEX	0.31	0.32	0.20	**0.40**

During experiments, the following variations of this method were tested:

- as described, with n-gram result list = 8—*ORIGINAl*,
- during analysis of the data, based on which the translation is being made, author found data errors, which may influence the overall score. The text errors were related to wrong UTF-8 character encoding, which gave wrong text for Latin characters in Eurlex corpus—*FIXED*,
- with using *Morfologik*[17] analyzer polish data—*MORFOLOGIK*,
- with extended with n-gram result list = 20—*EXTENDED*.

For testing purposes, two sets of 100 pairs of Polish-English phrases were created (Table 3):

- concerning every-day language—*COMMON*,
- concerning legal phrases—*LEX*.

For translation from Polish to English, the test gave following *BLEU* results (Table 4):

For translation from English to Polish, the test gave following *BLEU* results (Table 5):

There is a percentile of translations, which were measured incorrectly. This happens because translation reference does not take into consideration variations of

[17] http://morfologik.blogspot.com.

Table 5 BLEU score for legal phrases

	ORIGINAL	FIXED	MORFOLOGIK	EXTENDED
COMMON	0.33	0.33	0.31	**0.39**
LEX	0.21	0.22	0.21	**0.25**

Table 6 Example translations

Phrase	Translation	Reference translation
w zwiazku z	has to do with	with reference to
co do sekundy	to the second	to the moment
blisko lez	close tears	near tears
dobrze zgadnac	okay guess	guess right
moim zdaniem	in my opinion	to my mind

given phrase (The same phrase may be translated in different way, dependent on the context). Examples of false-negative translations (Table 6):

5 Application

As an end result, a web service (*REST*), which implement above process, has been created. The use of this web service is as follows:

Usage:

```
http://localhost:8080/translate-web/rs/translate/{INDEX}?
                      direction={DIRECTION}&phrase={PHRASE}

INDEX - selected lucene index [COMMON | LEX]
DIRECTION - translation direction [ PL_EN | EN_PL ]
PHRASE - input phrase to translate
```

Example:

```
http://localhost:8080/translate-web/rs/translate/COMMON?
                      direction=PL_EN&phrase=karty
```

As result, a *JSON* object is produced. It holds each step of the translation process, which with the help of user's knowledge, provides useful translation help.

Result:


```
{
   "translations":[
      {"translation":"karty","comment":"cards, and, card, you"},
      {"translation":"do","comment":"do, to, you, on"},
      {"translation":"gry","comment":"game, games, over, of"},
      {"translation":"karty do","comment":"playing cards"},
      {"translation":"do gry","comment":"back in, to play"},
      {"translation":"karty do gry","comment":"playing cards"}],
   "best":"playing cards",
   "valid":true,
   "message":null
 }
```


6 Conclusions

Presented solution cannot compete against currently working *SMT* solutions like Joshua and Moses. As it is described in Moses manual [5] for corpora of 200,000 pairs, system has been able to produce the result, which was scored up to 0.78 using *BLEU* metric. Similar example [4] based on *Euparl* corpora alone, produced up to 0.20 higher than the described solution. Although the simplicity of the process and small amount of resources necessary to aid a user to translate a phrase makes this approach legible and useful. Following data set has been prepared with on machine with configuration described in Sect. 2.1 in this chapter. Generated *Lucene* index consumed around 10 GB of disk space. The building process for around 42,000,000 pairs of text blocks took 3 h and 32 min. Approach which uses Moses [5] uses ten times more disk space and lasts up to 20 h for a model build. In order to improve the score, author suggest implementing simple language model for phrases. Therefore it is necessary split and align the longest text blocks used in this process.

References

1. Bond, F.: Machine translation introduction - lecture 1. NTT Communication Science Laboratories (2006)
2. Arnold, D.J., Balkan, L., Meijer, S., Humphreys, R.L., Sadler, L.: Machine Translation: An Introductory Guide. Blackwells-NCC, London (1994)
3. Papineni, K., Roukos, S., Ward, T., Zhu, W.J.: BLEU: a method for automatic evaluation of machine translation. In: Proceedings of the 40th Annual Meeting of the Association for Computational Linguistics (ACL), Philadelphia, pp. 311–318 (2002)

4. Koehn, P.: Europarl: a parallel corpus for statistical machine translation. In: MT Summit, vol. 5, pp. 79–86 (2005)
5. Machado, J.M., Hilario, L.F.: Moses for Mere Mortals. Tutorial. A machine translation chain for the real world (2014). https://github.com/jladcr/Moses-for-Mere-Mortals/blob/master/Tutorial.pdf

N-Gram Collection from a Large-Scale Corpus of Polish Internet

Szymon Roziewski, Wojciech Stokowiec and Antoni Sobkowicz

Abstract The paper is devoted to the processing of multi-Terabyte web archive. The aim of this work is to create N-gram collection based on a large-scale corpus of all Polish sites on the Internet provided by The Common Crawl Foundation project [1]. The data after lexical processing is used to extract flat N-grams compilations. They have many successful applications in machine learning within natural language processing.

Keywords Polish internet · N-grams · Large-scale corpus · Natural language processing · Cluster computing

1 Introduction

Fast growing Internet Website Collections covering almost all known subjects and languages, constitute an appealing resource for extraction of large-scale N-gram corpus of N-gram formation. Only a small fraction of Common Crawl general-purpose Internet crawl data set is devoted to Polish, which we focus on. There have been established widely recognized projects aiming to form web corpora, such as Common Crawl Corpus [1], WaCKy [2] and Google Books N-grams [3], etc. With recent improvement of computing power, storage availability and powerful, highly efficient scalable processing and computing frameworks, it has become feasible to build a large scale N-gram collection.

S. Roziewski (✉) · W. Stokowiec · A. Sobkowicz
Natural Language Processing Laboratory, National Information Processing Institute, Niepodlegosci 188 B Avenue, 00-608 Warsaw, Poland
e-mail: szymon.roziewski@opi.org.pl
URL: http://www.opi.org.pl/en/Natural-Language-Processing-Lab.html

W. Stokowiec
e-mail: wojciech.stokowiec@opi.org.pl

A. Sobkowicz
e-mail: antoni.sobkowicz@opi.org.pl

D. Ryżko et al. (eds.), *Machine Intelligence and Big Data in Industry*, Studies in Big Data 19, DOI 10.1007/978-3-319-30315-4_3

This paper describes the development of a large-scale Polish Internet linear N-gram Collection within National Information Processing Institute, currently consisting of 3 billion tokens which form more than 155 million sentences. The raw data size containing only Polish, originating from approximately 5 million Polish URLs, is nearly 10 GB. Lexical analysis has been performed and noisy data has been filtered.

The main goals of this work are to get practical experience of cutting-edge technologies commonly used for Big Data processing and N-grams corpora formation. We have created large-scale Polish N-gram collection based on Common Crawl Corpus i.e. unigrams, bigrams, trigrams, four-grams and five-grams.

2 Framework Infrastructure

Since we have focused on both the main goal and applications of modern effective technologies, we present more about data processing, chosen proper frameworks and scientific know-how. The whole process is divided into two fundamental phases: first stage consists of requesting for crawl data, transforming it and saving cleaned data into the Cassandra database. Second stage is related to N-grams formation on the basis of stored appropriately selected Polish Data Crawl. The former stage is relied on Akka framework written in Scala [4] which is powerful, highly effective and ideally fulfils needs related to distributed application, more details will be given in the further reading. Language detection is accomplished by using the Chromium Compact Language Detector package in version 2 (CLD2) [5]. Later, in Sect. 3, we will give more details about CLD2 and describe why we have chosen this library. In the second step we have used Apache Spark Framework [6] in order to process previously collected data and finally build the N-gram collection. The main logic of data processing and management is presented in Fig. 1.

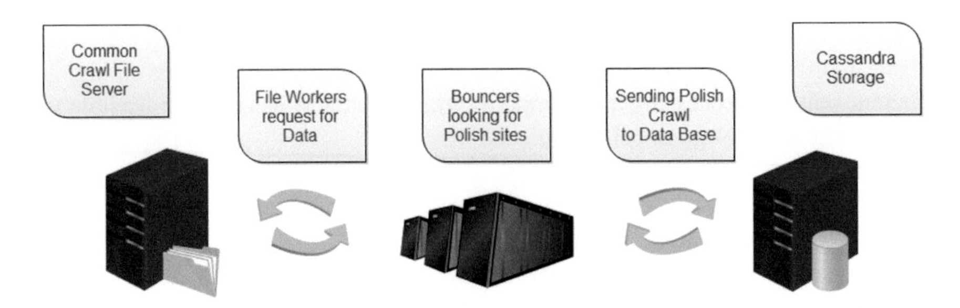

Fig. 1 The main logic for the first stage processing. Common Crawl Corpus is stored on the file servers. File Workers request for data dump and further Bouncers process the web documents for detection of Polish content by using CLD2 library. In the end, Polish Data Crawl is stored in the Cassandra data base

2.1 Dataset and Actors

January 2015 Crawl Archive, on which this study has been based, is approximately 140 TB in size and contains 1.82 billion web pages. Although, we have analysed only WET files,[1] which constitutes a fraction of it, around 7.5 TB. In order to facilitate data processing, the aforementioned dataset has been divided into 140 MB gz-compressed files containing numbers of parsed websites with extracted text and relevant meta-information.

Since textual information resides in disjoint files, it is straightforward to build N-gram Collection in parallel. Due to the fact, that processing web-scale data requires not only passing millions of messages concurrently, but also handling multiple failures (e.g. data store unavailability) we have decided to use Akka framework: a high-performing, resilient, actor-based runtime for managing concurrency.

Actors are objects which encapsulate state and behaviour, they communicate by exchanging messages. Theoretical underpinnings for actor model, a mathematical model of concurrent computation, are described in [7]. In Akka actors are organized in hierarchical structure known as actor system which is equipped with supervisor strategy, i.e. a straightforward mechanism for defining fault-tolerant service.

In our study actor system begins with Master Actor, an actor responsible for iterating over file that list WET files and dispatching their URL paths to individual File Workers. In an effort to avoid context-switching we have decided to limit the number of File Workers to the number of cores available in cluster on which program has been run. After receiving message with URL address each File Worker begins to download compressed file and by extracting plain text. Subsequently, each of the web site is sent to Bouncer Actor, which operates independently for language identification, and eventually for storage in Cassandra data base.

2.2 Cassandra Storage

Authors have chosen Apache Cassandra as data storage, which is very efficient distributed document database, developed by Apache Software foundation and which offers scaling support, thought to lessen read/write latency [8]. Apache Cassandra provides an automatic data replication between nodes and offers data model which is hybrid between both key-value pair and column-oriented models. There are several ready-to-use connectors for different programming languages, working with Akka, we have adopted connectivity drivers dedicated for Scala, developed by DataStax [9–11].

In our scenario we use two-node Cassandra Cluster with two tables. The first table is designed to store the list of processed WET files and the second one is for processed Polish Crawl. Keeping information about processed files during the run

[1]Files representing source code of crawled websites, containing tens of thousands web pages, consisting of special header with meta information and plain text as site content.

allow us to skip already analysed files in case of system crash. Processed text chunks stored in Cassandra enables us to perform further computations either from Spark application or any other non-distributed program.

3 Data Preparation

As we mentioned before, January 2015 Crawl Archive has been processed and that constitutes only a tiny fraction of Common Crawl Datastore which consist of petabytes of data collected over seven years [1]. This Crawl Archive contains over 25,000 segments, each of which is approximately 300 MB, what indicates that more than 7.5 TB of data has been analysed. The initial step was to download the data from Common Crawl Storage by Akka File Workers. This step turned out to be the most time and resource consuming. Afterwards, File Workers transmitted the chunks of text to Bouncer Actors, while passing through the crawl data file. These actors were to detect Polish content based on the first 800 bytes sent to CLD2 language detector.

The choice of language indicator was mainly motivated by its speed and accuracy [12], since language determination was one of the most time consuming process during gathering data crawl except to data downloading itself, and Chromium Compact Language Detector performed efficiently in both cases, and its accuracy has been even improved in version 2. The character encoding of text must be converted to valid UTF-8 charset before sending to CLD2. This language indicator returns the top three languages found and their probabilities, for multi-language web page. CLD2 is based on Naïve Bayesian classifier, implementing three different token algorithms in order to deal with particular languages in terms of language properties. In most cases, language distinction comes from quadgrams[2] scoring, for certain exotic languages the analysis is performed based on single letters scoring. Scoring process is done on lowercased Unicode letters and marks, after stripping digits, punctuation and HTML tags. The training corpus is constructed manually from selected web pages for each language, and then expanded automatically with precaution by scraping another 100 million websites. The table-driven, both extraction of letter sequences and scoring is vastly optimized in terms of speed and and space. It performs roughly 10 times faster than other language detectors and covers over 70 languages in 1.8M of x86 code and data structures. The main lookup table, covering 50 languages, consisting of quadgrams is 256 K four-byte items [5]. Indication performed over the average web page of 30 KB raw text with HTML tags takes only 1 msec on a current x86 processor. The CLD2 package can distinguish 83 languages, from all around the world.

Barely a tiny fraction of downloaded data was Polish, approximately 2.7 ‰. All aforementioned steps were performed in parallel with using the power of cluster as much as it was possible. Further processing relies on the Polish websites collected in Cassandra storage.

[2] Sequence of four adjacent letters.

Since HTML tags have been filtered out from WET files, we have proceeded to the next step. The text was split into sentences by using tokenizer from NLTK python toolkit [13]. NLTK package is a powerful and very effective platform for working with human language data. Since Apache Spark supports python language and NLTK is a very handy, well-developed tool, we have decided to take an advantage of it. The iteration over all documents within Polish Data Crawl mined from fast Cassandra datastore, were performed. During one cycle, each document has been tokenized into sentences and next, the iteration over these phrases occurred. In the meanwhile, words unfitting the template by the means of non-alphanumerical characters, numbers or non-Polish tokens were filtered off. For this purpose, we applied regular expression processor. Subsequently, back to the outer loop, N-gram sets were generated and stored in a hash map structure for ease counting reason.

The analysis of PDC contains nearly 5 million documents and forming around 155 million sentences consisting of 3 billion tokens, meaning that the most nested loop was performed as many as 3 billion times. Since the amount of work was massive, time consumption was substantial by the means of nearly 20 h computing.

4 Linear N-Grams

This section shows the results of building the N-gram corpora and gives more details of the scored collection, we will report on a few statistic features, show N-gram counts on the graph charts and give several N-gram examples for specific cases as well. After analysis made on the whole gathered Polish Crawl, we have constructed the N-gram Collection of the Polish Internet. The N-grams have been composed using fast python word processor running on Spark. Compilation of unigrams, bigrams, trigrams, tetragrams and pentagrams has been made. The first step to accomplish this task was to sort each set consisting of N-gram entry and its occurrences, for all the N-gram types with respect to N-gram counts, in descending order. Polish diacritic characters have been removed, whereas stop words have remained.

We give a few lists of N-gram examples, mainly we are focused on the first top ten entries, we also provide illustration from the middle of collection and objects at the end of it.

In Table 1 there are lists of unique unigrams, there are three segments, first is called top ten and relates to ten the most frequent unigrams, the second segment is from the middle of a set, and the last one is caught from the tail of the unigram collection. One can see that the most popular are neither nouns nor verbs, but prepositions and conjunctions, their occurrences are high and quite fast diminishing. In the second segments nouns and adjectives dominate. The last section is useless and unprofitable, it consists mainly from misprints or words fused together. But if we had removed the superfluous tail, we would have extracted more accurate unique unigrams.

In Table 2 we list samples of unique bigrams from three segments. The top ten bigrams are short common Internet phrases, their counts are decreasing a bit slower

Table 1 Overview of samples of unique unigrams and their counts

Top ten		2nd segment		3rd segment	
Sample	Count	Sample	Count	Sample	Count
w	74,129,969	drozdze	3833	gryangielska	7
i	59,263,299	trwalego	3833	orgowie	7
z	42,941,242	pchly	3833	gotowedomy	7
na	41,685,497	wieliczki	3833	formnie	7
do	34,425,979	polsko-niemiecki	3833	datasheecie	7
sie	24,398,982	polany	3833	krewetkariach	7
nie	23,983,941	koscielne	3833	jersli	7
to	20,904,222	silnikowego	3833	agrona	7
o	19,516,831	analityka	3833	liptowie	7
a	15,561,409	popielniczka	3833	nugdy	7

Table 2 Overview of samples of unique bigrams and their counts

Top ten		2nd segment		3rd segment	
Sample	Count	Sample	Count	Sample	Count
hotele, w	4,990,264	berlin, tajna	3402	pochlaniacza, co	4
w, poblizu	4,055,634	male, zuzycie	3402	zazwyczaj, koszyki	4
zaloguj, sie	1,945,524	nr, praca	3402	www, planet-poker	4
w, tym	1,720,763	srebro, prosze	3402	tylko, kopia	4
sie, w	1,401,364	pedzla, wykonany	3402	antropiczna, jest	4
prosze, o	1,261,377	publiczny, Krakow	3402	praca, wielka	4
sie, z	1,239,981	koszt, autobusu	3402	data, wrzesien	4
sie, na	1,162,140	kretowiny, gulbity	3402	hotele, maceio	4
napisz, recenzje	1,148,995	danym, lokalu	3402	wyrownawczy, plynu	4
nie, ma	1,135,745	zamowieniu, wagonow	3402	najmu, sprzedaly	4

than for unigrams. Occurrences from the second part are kind of idiomatic expressions. And the last segment is, as before, hardly practical.

The Table 3 illustrates trigrams. The most common are idiomatic expressions and the first is somehow associated with the most popular bigram, their counts decrease

Table 3 Overview of samples of unique trigrams and their counts

Top ten		2nd segment	
Sample	Count	Sample	Count
hotele, w, poblizu	3,406,762	trafnosci, od, najnowszych	3406
z, o, o	2,952,232	postow, nie, mozesz	3406
lovciam, cie, justinek	955,962	witam, wszystkich, serdecznie	3406
justinek, lovciam, cie	955,476	a, jaki, jest	3406
cie, justinek, lovciam	955,476	prosimy, o, zapoznanie	3406
przejdz, do, sklepu	705,884	jedynki, kreacje, wywiady	3406
windows, internet, explorer	541,049	rozpowszechnianie, niniejszego, artykulu	3406
internet, explorer, mozilla	539,598	stambul, stambul, last	3406
explorer, mozilla, firefox	539,598	zdrowia, i, choroby	3406
moze, nie, byc	539,597	napoje, herbata, kawa	3406

Table 4 Overview of samples of unique four-grams and their counts

Top ten	
Sample	Count
Internet, explorer, mozilla, firefox	530,364
tripadvisor, zauwazylismy, ze, uzywasz	522,089
zauwazylismy, ze, uzywasz, nieobslugiwanej	522,089
ze, uzywasz, nieobslugiwanej, przegladarki	522,088
tripadvisor, moze, nie, by	522,087
moze, nie, byc, wywietlana	522,087
witryna, tripadvisor, moze, nie	522,087
nie, byc, wyswietlana, poprawnie	522,087
obslugiwane, sa, nastpujace, przegladarki	522,086
sa, nastepujace, przegladarki, windows	522,085

calmly. The second segment is very idiomatic. Third section is not printed out this time, since it represents negligible data.

We present only top four- and five-grams in Tables 4 and 5, respectively. It is clear that all top four-gram entries are somehow related to each other. They form quite logic sentence when read consecutively. Their occurrences are now almost constant without much diversities. The very same conclusions came up when analysing top frequent five-grams shown in Table 5, with one exception that the first two entries

Table 5 Overview of samples of unique five-grams and their counts

Top ten	
Sample	Count
ograniczona, odpowiedzialnosci, krsodpis, z, krs	667,764
z, ograniczona, odpowiedzialnoscia, krsodpis, z	667,764
tripadvisor, zauwazylismy, ze, uzywasz, nieobslugiwanej	580,895
zauwazylismy, ze, uzywasz, nieobslugiwanej, przegladarki	580,894
tripadvisor, moze, nie, byc, wyswietlana	580,893
witryna, tripadvisor, moze, nie, byc	580,893
moze, nie, byc, wyswietlana, poprawnie	580,893
bslugiwane, sa, nastepujace, przegladarki, windows	579,919
nastepujace, przegladarki, windows, internet, explorer	578,946
sa, nastepujace, przegladarki, windows, internet	578,946

Table 6 Overview of total numbers occurrences for all type N-grams and their weights

N-gram type	Total # occurrences	Collection weight (GB)
Unigram	2,972,108,905	3.9
Bigram	2,836,982,256	29
Trigram	2,742,722,499	53
Four-gram	2,792,853,174	70
Five-gram	2,324,914,403	84
Total	13,669,581,237	245

are not correlated with the rest. We can see two subsets of five-grams forming the top ten collection sample.

We present some characteristics of N-gram collection, in Table 6, we state the total number of occurrences for each N-gram type and related storage weight. There is quite a lot of data in total, which is formed out of not aggregated N-grams. There is almost 3 billion unigrams in total what as it has been mentioned in the previous section. Number of bigrams are a little bit smaller, in general collections for each N-gram type decrease in size with larger N, what is in agreement with [14]. The Table 6 shows that there is plenty of data to be merged in order to obtain collections of unique N-grams.

We have aggregated the previous collections (see Table 6), and the result is attached in Table 7, consisting of unique N-grams. Now, the size of data highly diminished by 99.4 %. Now, the result shows a little bit different characteristics, the number of bigrams is fairly smaller than trigram occurrences which is in contrast with the disjoint collection from Table 6.

As a last step, we need to reduce the N-gram set by confusing N-grams which are mainly misprints and fused words. All in all, we have a Polish N-gram Collection which can be seen in Table 8. The size of data is further decreased, on average by the

Table 7 Overview of total numbers of unique N-grams and their weights

N-gram type	Total # occurrences	Collection weight (MB)
Unigram	29,758,516	529
Bigram	4,795,792	106
Trigram	7,278,725	209
Four-gram	8,080,472	292
Five-gram	7,439,693	326
Total	57,353,198	1460

Table 8 Overview of total numbers of unique N-grams and their weights after truncating of misprints and fused words

N-gram type	Total # occurrences	Collection weight (MB)
Unigram	2,985,800	50
Bigram	2,608,100	60
Trigram	3,790,700	113
Four-gram	4,277,000	159
Five-gram	3,617,000	163
Total	17,278,600	545

Table 9 Statistics of unique N-grams' lengths

N-gram type	Mean	SE	Median	10th PCTL	90th PCTL
Unigram	8.79	3.07	9	5	13
Bigram	15.35	4.61	15	10	21
Trigram	22.32	6.03	22	15	30
Four-gram	30.03	7.33	29	21	39
Five-gram	38.34	8.46	38	28	49

fraction 0.63. The collections shown in Table 7 consists of the words that occur at least 10 times [14]. The obtained result is comparable by the means of essence with the collection in Table 7, it is much smaller in size though.

In Table 9 one can find several statistics concerning N-gram length for each N-gram type i.e. mean, standard error, median and two percentiles: 10th, 90th. Our research infers that the mean of Polish word length is 8.79 with standard deviation 3.07, what seems to be a little bit overestimated, mainly because fused words can be found in the N-gram corpus for unigrams, which should be somehow detected and removed from the corpora. It means that not only the mean might be affected but all statistics more or less. Statistics for other N-gram types may be altered for the same reason.

In Fig. 2 we show N-gram occurrences for each N-gram type with respect to its length. The data has been aggregated appropriately, thus all counts are summed up for each N-gram length. One can see a few phenomena: number of occurrences have the same order of magnitude; for higher N-gram orders the curve shapes are bell-

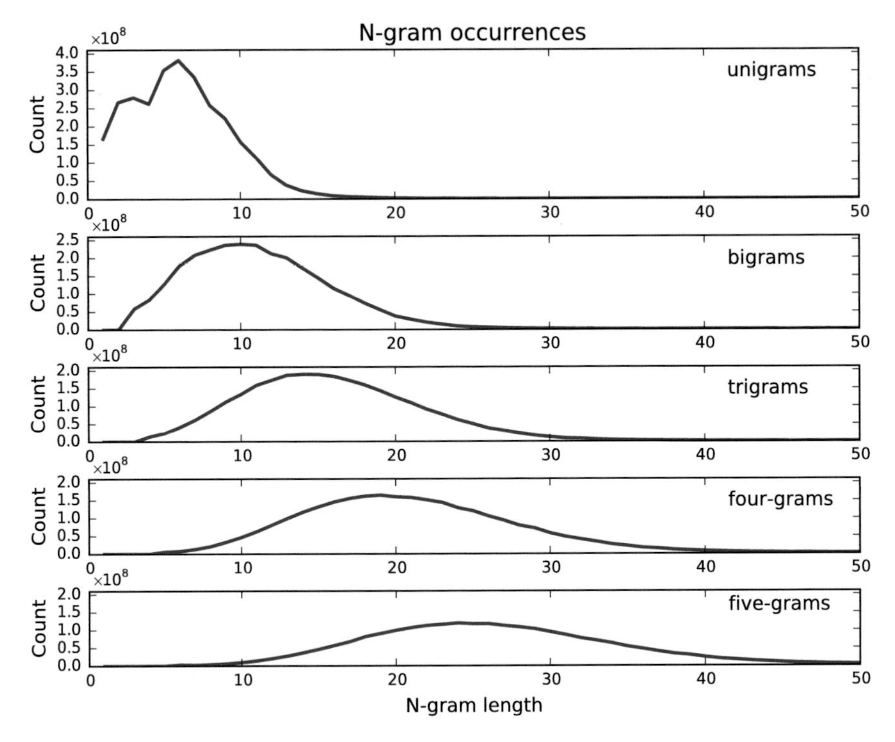

Fig. 2 The N-gram curves show counts of each N-gram type with respect to its length by the means of characters, and illustrate the data from Table 8. The line chart for unigrams shows two maxima, the first one is related to stop words and the second one is for the most frequent size of Polish word which is 6. The curves of N-gram occurrence functions for bigrams, trigrams, four- and five-grams are right-skewed bell-curve-like, with long tails widening with increasing N-gram length. The chart curvatures lessen with increasing N-gram rank. The maxima are right-shifted by the value of 5

curve-like, widening with increasing N-gram rank; with increasing N-gram order it is meaningful that the charts are getting more right-skewed, the curve maximums are more right-shifted and the curvature of charts lessen; the mean and median are shifted to growing N-gram length, in particular for greater order N-grams, which is caused by the long right tail of curves. The elongated tails for large N-gram length may occur because of words concatenation, which has not been detected. The shape of the unigram curve is particularly different by the means of two local maxima. The first maximum may come from the stop words which have not been stripped, and the second one is for the most popular word length which is 6 with its occurrence equal nearly 400 mln. We can see that the maxima are shifted approximately by the most frequent word unigram size. We may infer from this fact that the average size of Polish word is roughly 6.

In the end we present summary of our study in the graph charts (see the Fig. 3). The line charts are prepared for the top unique thirty N-grams for each N-gram type, which show N-gram occurrences. It is clear that there are two main line characteristics, the first is for three curves describing unigrams, bigrams and trigrams, and the second

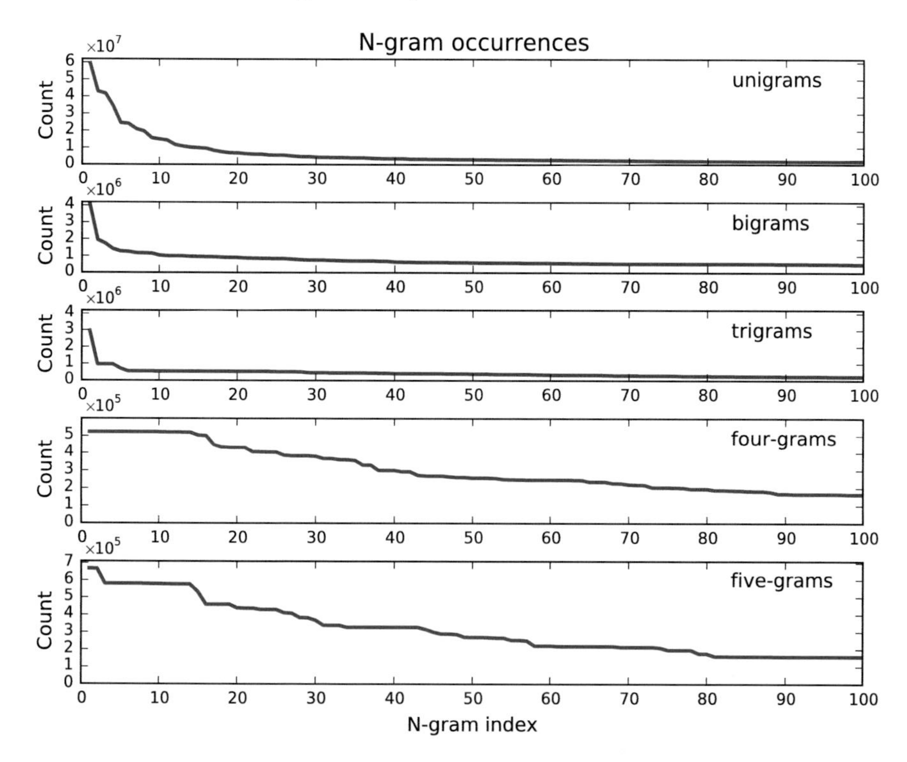

Fig. 3 The N-gram charts summarize occurrences each of N-gram type for the most frequent 100 N-grams. The line charts show changes in N-gram counts from the N-gram frequency table. The shapes of N-gram occurrence functions are more steep for unigrams, bigrams and trigrams, whereas for four- and five-grams the curvature is more flat, with long tails widening to the end of the N-gram collection. The general trend is asymptotic downwards for the first two charts and constancy for the other

for four- and five-grams. One can notice that first-group lines decline asymptotically, which is a kind of Zipf's Law [15] manifest. This aforementioned law states that the occurrence of a word in a corpora is inversely proportional to its rank in the frequency list. The curves for four- and five-grams are more flat and constant, which comes from the fact that N-grams in these groups are more correlated to each other among a group, what is clearly shown in Tables 4 and 5. Occurrences of tripadvisor related phrases in larger order N-grams may come from quirks of Common Crawl crawler algorithm.

5 Conclusions and Future Work

We have presented our research on the Polish Internet Corpus based on the Common Crawl January 2015 Archive. After analysing terabytes of data, we have created rich N-gram collections for five N-gram types. The main characteristics with several statistic measures have been shown. The most frequent unigrams are stop words, in the area where the Zipf'a curve is less steep, one can find more informative unigram entries. The tail of the collection is built from unworthy items, and have been pruned. Unigrams, bigrams and trigrams show pretty similar behaviour in occurrence, and for four- five-grams the same is met. The counting curves with respect to N-gram index in frequency table for four- and five-grams are quite flat, which can be explained by N-gram correlations. We have estimated the average length of Polish word which is approximately 6. Another remark is that a lot of duplicated data found on the Internet, which can be inferred from the same misprints, fused words and data aggregation. It is tempting to exclude concatenated words and improve handling with character coding for better accumulation. Having built such an infrastructure allows for additional analysis of Common Crawl Archives to enrich Polish N-gram Collection.

Acknowledgments The authors would like to thank the Common Crawl Foundation for the data free availability.

References

1. Common Crawl Foundation. http://commoncrawl.org
2. Baroni, M., Bernardini, S., Ferraresi, A., Zanchetta, E.: The WaCKy wide web: a collection of very large linguistically processed web-crawled corpora. Language resources and evaluation, vol. 43(3), pp. 209–226. Springer, New York (2009)
3. Google Books N-grams. http://books.google.com/ngrams
4. Akka 2.3.10 for Scala 2.10. http://akka.io
5. Chromium Compact Language Detector 2. http://code.google.com/p/cld2
6. Apache Spark. http://spark.apache.org
7. Hewitt, C., Bishop, P., Steiger, R.: A universal modular actor formalism for artificial intelligence, IJCAI (1973)
8. Rabl, T., Sadoghi, M., Jacobsen, H.A., Gomez-Villamor, S., Muntes-Mulero, V., Mankovskii, S.: Solving big data challenges for enterprise application performance management, VLDB (2012)
9. http://github.com/datastax/spark-cassandra-connector
10. http://github.com/datastax/java-driver
11. http://github.com/datastax/python-driver
12. Kanerva, J., Luotolahti, J., Laippala, V., Ginter, F.: Syntactic N-gram Collection from a Large-Scale Corpus of Internet Finnish. IOS Press, Amsterdam (2014)
13. Natural Language Toolkit 3.0. http://www.nltk.org
14. Yang, S., Zhu, H., Apostoli, A., Cao, P.: N-gram statistics in English and Chinese: similarities and differences, Google (year)
15. Powers, D.M.W.: Applications and Explanations of Zipf's Law. Association for Computational Linguistics, Stroudsburg (1998)

Study Fields Clustering Using KRK Competences

Marek Kozlowski

Abstract The paper refers to the topic of study fields clustering using extracted information from semi-structured documents, namely documents describing study field's KRK competences. KRK competences are the specialized descriptions of the qualifications, which students gain after graduation from the given study field. The proposed method enables extracting and processing KRK competences from diverse types of semi-structured documents. It consists of two stages: (1) entity extraction from documents (building vectors of KRK competences for each study field), and (2) study fields clustering using those competence representations. Polish KRK competence files, describing almost 3000 study fields in Poland, were used as a corpora. The method and its stages are thoroughly analyzed. The results allow to compare and identify similar study fields according to theirs final effects of education.

1 Introduction

We are living in times of exponential growth of the production of unstructured data, such as web pages, social media, documents etc. Such mass of information is unlikely to be analyzed by humans, so there is a strong drive to develop automatic methods capable to retrieve knowledge from unstructured data.

Since a decade, text clustering has become an active field of research in the machine learning community. Most of the approaches are based on the term occurrence frequency. The performance of such surface-based methods can decrease when the texts are too complex, i.e., ambiguous. One alternative is to use the semantic-based approaches to process textual documents according to their meaning. Furthermore, research in text categorization has been focused on flat texts whereas many documents are now semi-structured format.

In this paper we present the application of diverse clustering methods in order to discover knowledge from the repository of semi-structured documents (describing

M. Kozlowski (✉)
National Information Processing Institute, Warsaw, Poland
e-mail: mkozlowski@opi.org.pl
URL: http://www.opi.org.pl

D. Ryżko et al. (eds.), *Machine Intelligence and Big Data in Industry*,
Studies in Big Data 19, DOI 10.1007/978-3-319-30315-4_4

KRK competences). The proposed method exploits extensively information extraction methods (regular expressions) and information retrieval methods (flat clustering, hierarchical agglomerative clustering, patterns mining and grouping them into the lattice). Our approach is evaluated on the unique dataset describing KRK education competences of 2800 study fields in Polish Higher Education System.

2 Preface

Since the late 1990s, the Bologna Process has offered an incentive for many European countries to reform their educational systems and to make them more comparable and explicit. Nowadays, 46 European countries are involved in the process and one of its outcomes is the European Qualifications Framework (EQF), the system of explicit and comparable qualifications.

The EQF aims to establish a common reference framework as a translation device between different qualification systems and their levels. This framework comprises general, higher and vocational education and training, and should lead to better transparency, comparability and portability of citizen's qualifications (e.g. diplomas, certificates etc.). The Qualification Framework strengthens competitiveness by enhancing mobility between the European educational systems and the labour market. Individuals and employers will be able to use the EQF for better understanding and comparing the qualifications levels of different countries.

Within the framework of the Bologna Process, the European ministers of education have agreed that each country should develop it's own framework for degrees and qualifications (qualifications framework). Polish framework is called Krajowe Ramy Kwalifikacji (KRK competence framework). KRK competences are specialized descriptions of the abilities/qualifications, which students gain after graduation from the given study field. In Polish Higher Education System each study field should have the KRK competences (described in the semi-structured file) assigned to it. It means that study fields are linked to the files (pdf/doc), where in a table or in an enumeration form the KRK competences are placed. Polish KRK competence corpora consists of 3550 files and refers to about 2800 study fields.

3 Related Work

This paper refers to the two topics: information extraction (semi-structured extraction) and information retrieval (diverse clustering methods).

Information extraction (IE) is the task of automatically extracting structured information from unstructured or semi-structured machine-readable documents. The most popular tasks in IE are: named entity recognition, co-reference and relationship identification, table extraction, or the terminology extraction. There are various methods for information extraction like regular expressions, decision trees, Bayesian methods or Conditional Random Fields. In the paper [1] there are presented various IE methods in order to extract scientific profiles from web pages.

Document clustering is the form of unsupervised learning, which groups a set of documents into subsets called clusters. The algorithm's goal is to create clusters that are internally coherent, but clearly different from each other. In other words, documents within a cluster should be as similar as possible, and documents in one cluster should be as dissimilar as possible from documents in other clusters [11]. The goal of text clustering in information retrieval is to discover groups of semantically related documents. At the root of the clustering documents lies van Rijsbergens [14] cluster hypothesis: closely associated documents tend to be relevant to the same requests, whereas documents concerning different meanings of the input query are expected to belong to different clusters.

The key input to a clustering algorithm is the distance measure. Two major classes of distance measure are: Euclidean, Non-Euclidean. A Euclidean space has some number of real-valued dimensions and it is based on the locations of points in such a space. A Non-Euclidean distance is based on properties of points, but not their location in a space. Euclidean distance is: L_1 norm (like Manhattan distance), or L_2 norm (most common notion of distance). Non-euclidean distances are: Jaccard distance, Cosine distance or Edit distance (number of inserts and deleted to change one string into another).

Clustering methods are usually divided into two groups: flat clustering and hierarchical clustering. Flat approach creates a flat set of clusters without any explicit structure that would relate clusters to each other. Hierarchical clustering creates a hierarchy, a structure that is more informative than the unstructured set of clusters. These features of hierarchical clustering come at the cost of lower efficiency. In our applications we used both of those methods.

Approaches to text clustering can be also classified as data-centric or description-centric [3].

The data-centric approach focuses more on the problem of data clustering, rather than presenting the results to the user. Scatter/Gather [4] is an example, which divides the dataset into a small number of clusters and, after the selection of a group, performs clustering again and proceeds iteratively using the Buckshot-fractionation algorithm. Other data-centric methods use hierarchical agglomerative clustering [10] that replaces single terms with lexical affinities (2-grams of words) as features, or exploit link information [17].

Description-centric approaches are more focused on the description that is produced for each cluster of documents. This problem is also called descriptive clustering: discovery of diverse groups of semantically related documents associated with meaningful, comprehensible and compact text labels. Accurate and concise cluster descriptions (labels) let the user search through the collection's content faster and are essential for various browsing interfaces. The task of creating descriptive, sensible cluster labels is difficult—typical text clustering algorithms rely on samples of keywords for describing discovered clusters. Among the most popular and successful approaches are phrase-based, which form clusters based on recurring phrases instead of numerical frequencies of isolated terms. STC algorithm employs frequently recurring phrases as both document similarity feature and final cluster description [16]. KeySRC improved STC approach by adding part-of-speech pruning and dynamic selection of the cut-off level of the clustering dendrogram [2]. Description-Comes-First (DCF) approach was introduced in the work [13] as an algorithm called Lingo. Description-Comes-First is a special case of description-centric approach, it first attempts to find good, conceptually varied cluster labels and then assign documents to the labels to form groups. The Lingo algorithm combines common phrase discovery and latent semantic indexing techniques to separate search results into meaningful groups. Phrase-based methods usually provide good results. They report some problems, when one topic is dominating. Navigli and Crisafulli, Di Marco [5, 6, 12] present a novel approach to snippet clustering, based on the automatic discovery of word senses from raw text. The proposed method clusters snippets based on their semantic similarity to the induced query senses.

Inspired by the above description-centric algorithms we introduced a novel method for clustering web search results based on frequent termsets mining [8]. First, we acquire the senses of a query by means of a word sense induction method that identifies meanings as trees of closed frequent termsets. Then we cluster the search results based on their lexical and semantic intersection with induced senses. We do not use any external corpora, the sense induction is performed only on the search results. The search results are distributed among matching senses. Finally we also use some diversification techniques in order to rerank clusters and theirs content. This method [8] has been used in clustering of KRK competences and corresponding to them study fields.

4 Approach

In order to perform computations we have built dedicated tools crucial for the KRK competence symbols extraction, clustering study fields according to theirs KRK representations, or identifying KRK patterns and structurizing them into sub-trees.

4.1 Extraction of KRK Competences

Extraction of KRK Compentences concerns processing the files describing each study field with the list of KRK competences. The sample file, containing KRK competence symbols, is presented in Fig. 1. There are mainly pdf/doc files containing the tables or list of symbols of competences.

Kierunek *edukacja artystyczna w zakresie sztuk plastycznych* sytuuje się w Obszarze kształcenia w zakresie sztuki i realizuje wszystkie efekty kształcenia w tym obszarze. Absolwent kierunku posiada wiedzę oraz umiejętności wykwalifikowanego plastyka – twórcy w zakresie sztuk plastycznych oraz uczestnika i animatora kultury współczesnej. Jest przygotowany do podjęcia i realizacji różnorodnych przekazów wizualnych i medialnych do celów artystycznych, edukacyjnych, poznawczych i użytkowych. Przygotowany jest do podjęcia pracy zawodowej w: ośrodkach i instytucjach kultury, sztuki i edukacji pozaszkolnej; mass-mediach; strukturach promocyjnych i reklamowych oraz szkolnictwie – po ukończeniu specjalności nauczycielskiej.

2. Efekty kształcenia

Objaśnienie oznaczeń:

K (przed podkreślnikiem)	– kierunkowe efekty kształcenia
W	– kategoria wiedzy w efektach kształcenia
U	– kategoria umiejętności w efektach kształcenia
K (po podkreślniku)	– kategoria kompetencji społecznych
01, 02, 03 i kolejne	– numer efektu kształcenia

Symbol	**Efekty kształcenia dla kierunku studiów** *edukacja artystyczna w zakresie sztuk plastycznych* **Po ukończeniu studiów** pierwszego stopnia **na kierunku studiów** *edukacja artystyczna w zakresie sztuk plastycznych* **absolwent:**	**Odniesienie do efektów kształcenia w obszarze kształcenia w zakresie** *sztuki*
	WIEDZA	
	Wiedza o realizacji prac artystycznych	
K_W01	Formułuje pojęcia i terminy związane z dyscyplinami artystycznymi	A1_W10
K_W02	Objaśnia budowę dzieła plastycznego: rodzaje i zasady kompozycji na płaszczyźnie i w przestrzeni	A1_W10
K_W03	Wskazuje środki ekspresji i umiejętności warsztatowe pokrewnych dyscyplin artystycznych	A1_W10
	Wiedza i rozumienie kontekstu sztuk plastycznych	
K_W04	Tłumaczy podstawowe linie rozwojowe w dziejach poszczególnych dyscyplin artystycznych i zna publikacje na ten temat	A1_W11
K_W05	Zna podstawowe style w sztuce ich tradycje twórcze i chronologię	A1_W12
K_W06	Tłumaczy konteksty kulturowe, społeczne, religijne i polityczne sztuki i rozpoznaje związki pomiędzy dziełem sztuki a refleksją teoretyczną i myślą estetyczną na	A1_W11 A1_W12

1

Fig. 1 Sample file describing study field 'edukacja artystyczna' with KRK competences placed in a table

The information extraction is performed as follows:

1. File processing using Apache Tika;
2. KRK competence symbols extraction using regular expressions;

File processing is done in order to fetch a content of file in a text format. Apache Tika toolkit detects and extracts metadata and text (file's content) from over a thousand different file types (such PPT, XLS, DOC(X), PDF). All of those files are parsed through a single interface, making Tika crucial for search engine indexing and content extraction. In our experiments we used the AutoDetectParser class in order to works with diverse types of files (xls, pdf, doc).

KRK competences are extracted from the retrieved file's content using well-defined regular expressions. Analyzing the KRK symbols we discovered some regularities in the notation, which give us possibility to define the one general regular expression as:

$$((\backslash p\{Upper\})((\backslash w\{1,3\})\_(\backslash w\{3\}))|(A\_(\backslash w\{3\})))$$

Summarizing, for each document we build the vector of KRK competence symbols. For the study field Geography at Poznan University the vector of retrieved competences counts 62 symbols and looks like $< S2A\_U08, S2A\_U09, \ldots S2A\_W07, \ldots, S2A\_K05, \ldots, P2A\_K01 >$. Such vector representation of competences is used in order to cluster the study fields.

4.2 Flat and Hierarchical Clustering

Flat clustering is efficient and conceptually simple, but requires a priori defined number of clusters as input and are usually nondeterministic. Hierarchical clustering creates a structure containing the history of grouping that presents in each iteration the state of clustering. Such representation is more informative than the unstructured set of clusters. In some applications (as finding similar study fields in the KRK vector space) we want a partition of disjoint clusters just as in flat clustering. In those cases the hierarchy needs to be cut at some point. There are two ways: to prespecify the number of clusters, or to prespecify level of similarity (minimum similarity between clusters). Analyzing the hierarchy of clusters is much simpler to decide about the cut off level (in our case the minimal similarity that enables to merge two clusters), which is acceptable for us.

Hierarchical clustering is either top-down or bottom-up. Top-down clustering proceeds by splitting initial cluster (containing all instances) recursively until singletons (individual instances) are reached. Bottom-up approach treats each document as a singleton cluster at the initial step and then merge them until all clusters are grouped into a single cluster containing all instances. This approach is called hierarchical agglomerative clustering, or shortly HAC. HAC is more frequently used in IR than top-down clustering so we also used the agglomerative approach in ours applications. HAC algorithm can be used with the different similarity measures as: single-linkage, complete-linkage, or average-linkage. Ours initial experiments prove that single linkage provides better results than complete-linkage, joining clusters based on the similarity of their most similar members. Complete-linkage clustering suffers from a special problem of KRK competence vectors, namely it pays too much attention to outliers, study fields that do not fit well into the global structure of the cluster.

We performed two clusterings in ours experiments: (1) flat deterministic clustering for the mirrors discovery, (2) HAC clustering for the similar study fields discovery.

4.3 SnS-Based Clustering

We have also applied a dedicated WSI-based web search result clustering method called SnSRC [8]. SnSRC consists of the following four steps: (1) preprocessing of results (transforming into a bag of words), (2) word sense induction method, (3) clustering of results, and (4) cluster sorting.

In the first step for a given query the interesting documents (snippets) are retrieved, and then iteratively processed into bag-of-words representation. In our case the bag-of-words representation is given initially as the vector of KRK competence symbols (each KRK competence symbol can be treated as a normalized word). Steps 2, 3 are performed as in the original version according to clustering snippets. The last step 4 is omitted. Some brief description of steps 2 and 3 is the next paragraphs.

Word sense induction step (2) is performed with the use of SnS [7, 9]. It is a word sense induction algorithm based on closed frequent sets and multi-level sense representation. SnS is a knowledge-poor approach, which means it does not need any kind of structured knowledge base about senses as well as the algorithms that have embedded deep language knowledge. Senses induced by SnS characterize better readability (are more intuitive), mainly because SnS discovers a hierarchy of senses showing important relationships between them. In other words the proposed method creates structure of senses, where coarse-grained senses contain related sub-senses (fine-grained senses), rather than flat list of concepts.

In our case the customized SnS algorithm consists of three phases, which we present below.

In Phase I KRK patterns are discovered from KRK competence vectors describing study fields. The patterns are closed frequent sets in the KRK competence space. The KRK competence vectors are treated as transactions (itemsets are replaced by sets of KRK symbols) and the process of mining closed frequent sets is performed with the use of the CHARM algorithm [15].

Phase II is devoted to forming KRK patterns into sense frames, building a hierarchical structure of senses. In some exceptional states few sense frames may refer to one sense, it may result from the corpus limitations (lack of representativeness and high synonymity against descriptive terms).

In Phase III, sense frames are clustered. The clusters of sense frames are called senses.

The clustering step (3) is performed in two phases: first, simultaneously during sense induction, and then after sense discovery for those results that remained not grouped. The first phase is based on the process of frequent set mining. Discovered closed frequent sets have support, and list of results, in which they appear. Senses are clustered sense frames. Each sense frame has the main pattern, so according to sense frames the results (study fields) containing the main pattern are grouped in the corresponding result cluster. Summarizing, for each sense a corresponding cluster of study fields is constructed. Let us note that after this phase is completed, there may remain study fields which are non-grouped. In the second phase, non-grouped instances are tested iteratively against each of the induced sense, and clustered to the closest senses.

4.4 Study Fields Search Engine

The search engine of study fields has been built, which goal is to find similar study fields to the given one using KRK competence vectors. In the proposed approach the closest study fields in the KRK vector space are presented in a flat list, where each item is also tagged with a number of common KRK competences. Figure 2 presents the top 10 results for the study field 'Coaching medyczny' (Table 1).

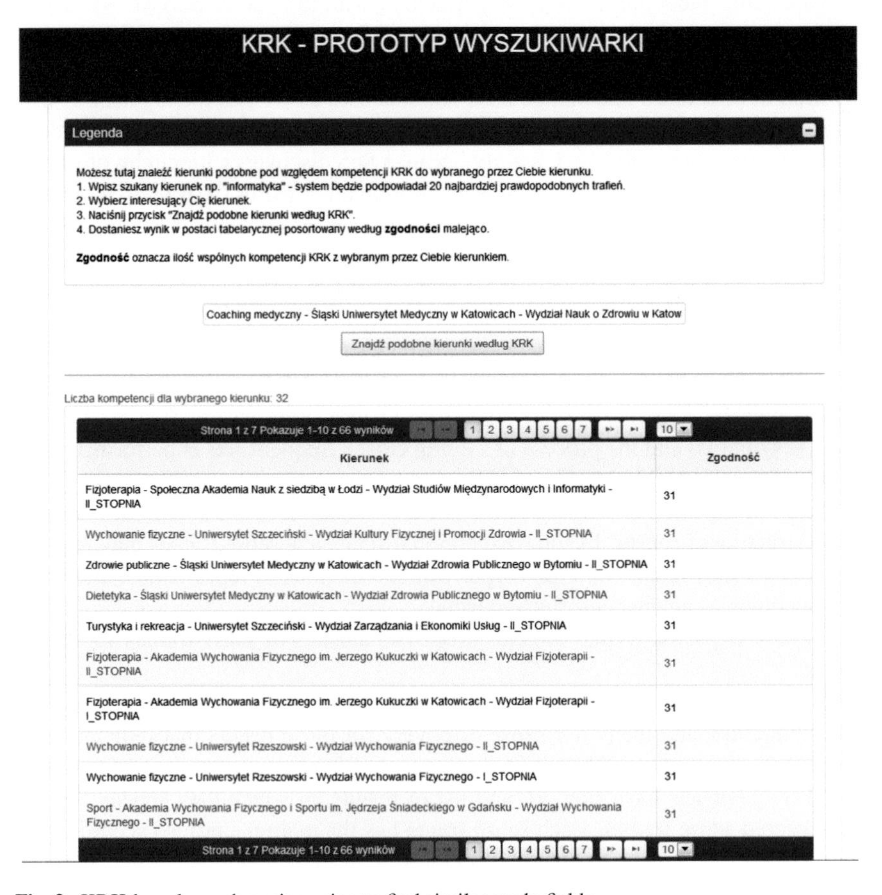

Fig. 2 KRK-based search engine using to find similar study fields

Table 1 Sample clusters of study fields having the same KRK competences (mirrors)

Cluster numbers	Study fields
1 (Size: 7)	**Politologia** (2550), **Stosunki miedzynarodowe** (2544), Fizyka techniczna (2606), Fizyka (2604), Matematyka (2598), Informatyka (2594), Astronomia (2601)
2 (Size: 4)	Wzornictwo (2), Grafika (3789), Malarstwo (6804), Architektura wnetrz (3790)
3 (Size: 4)	Rolnictwo (4225), Zootechnika (4022), Jakosc i bezpieczenstwo zywnosci (11153), Weterynaria (8007)
4 (Size: 4)	**Jazz i muzyka estradowa** (195), Bezpieczenstwo wewnetrzne (211) Administracja (6255), Bezpieczenstwo narodowe (140)
5 (Size: 2)	**Weterynaria** (4202), Architektura (406)

5 Experiments

5.1 Experimental Setup

Test sets. We conducted our experiments on the corpora created using data from the POLon system (the central system of Polish Higher Education). Below some details about this dedicated corpora:

1. 3550 files assigned to study fields of all universities;
2. average 1,3 file per study field;
3. 2700 study fields have assigned the KRK file(s), but only for 2200 of them assigned files are automatically processed i.e. we are able to automatically extract KRK symbols from them.

5.2 Results

We performed computations using all approaches described above, starting from the flat and hierarchical clustering in order to find clusters of equal or similar study fields according to KRK vector space.

Mirrors discovery consists in finding cluster of study fields having the same KRK competences (100 % similarity between sets of KRK competence symbols). This process is performed in order to find obvious anomalies in the data. There were discovered 150 clusters of mirrored competences (average size of a cluster is 11), some of them are presented below (the anomalies are in bold). Each study field is tagged with the POLon id (database id, which disambiguates study field, because e.g. political science can be taught both at Poznan University and Warsaw University).

We performed a second experiment aimed at clustering study fields using HAC algorithm. HAC clustering algorithm exploits single-linkage with cut-off = 0.7 and dedicated distance measure between KRK competence vectors. KRK dedicated similarity measure ($krkSim$) is defined by non-euclidean measure some way similar to Jaccard measure. Given a dictionary D of all KRK competence symbols and two KRK vectors (set of KRK symbol competences) $W_i \subseteq D$ and $W_j \subseteq D$ then the measure is expressed by the formula:

$$krkSim(W_i, W_j) = \frac{c(W_i \cap W_j)}{max(c(W_i), c(W_j))}$$

Here, $c(X_i)$ is a cardinality of vector X_i (vector of KRK symbol competences). Two KRK vectors are semantically close if the similarity measure $krkSim$ is higher than the user-defined threshold (e.g. 70 %). We should emphasize that the proposed measure is always lower than cosine measure built upon vectors, where each KRK symbol is a dimension (values are boolean). Comparing ours distance measure with the cosine measure the numerators are equal, but denominator $max(c(W_i), c(W_j))$ is always greater or equal than the cosine denominator. Experiments with cosine

Table 2 Sample clusters of study fields clustered by HAC (cut-off = 0.7)

Cluster numbers	Study fields
1 (Size: 4)	Geografia (1410), Turystyka i rekreacja (7664), Pedagogika (1700), Zarzadzanie srodowiskiem (7183)
2 (Size: 39)	Matematyka (3823), Astronomia (1479), Fizyka (4708), Chemia (5042), Informatyka (2788), Biofizyka (1484),........, Matematyka komputerowa (7806), Energetyka i chemia jadrowa (7962)
3 (Size: 324)	Socjologia (4497), Kulturoznastwo (2008), Nauki o rodzinie (1973), Historia sztuki (1725), Etyka (7747),......., Praca socjalna (1083), Profilaktyka spoleczna i resocjalizacja (1084)
4 (Size: 10)	Architektura wnetrz (3263), Malarstwo (6645), Grafika (2758) Edukacja artystyczna (1453),......., Sztuki wizualne (2761)
5 (Size: 15)	Coaching medyczny (10973), Sport (4763), Dietetyka (1463) Kosmetologia (6989), Fizjoterapia (7392),......., Analityka medyczna (6357), Wychowanie fizyczne (4822)

measure showed that some study fields are clustered too easy, therefore we decided to change this condition into more strict. The performed HAC clustering brings us 60 clusters (diverse distribution from 2 to 1000 instances into one cluster). The Table 2 shows sample clusters retrieved by HAC.

Apart from flat and HAC clustering we applied also the SnSRC algorithm, which can be treated as a clustering of patterns discovered in the KRK vectors. The SnSRC representation enables us to analyze some regularities in the KRK competences, and its structure itself (core competences, and some most popular extensions). Given minimum support 10 % we have discovered 80 000 patterns (closed frequent sets), which formed almost 4000 sense frames (trees clustering patterns). Sense frame is a multi-hierarchical structure organizing patterns. The root is a main contextual pattern, which is a representative label for the sense frame. The main pattern has subtrees, which are sub-pattern trees. Sub-patterns are supersets of the main pattern, but also they can be in subset relation among themselves, which brings to multi-level sub-pattern trees. Sense frames enable to build compact, concise representation of discovered patterns. The Fig. 3 shows some patterns clustered into the form of subtrees. The legend of some KRK competence symbols, used in the Fig. 3, is placed below.

1. $S1A\_W02$—graduate person has knowledge about social structures (also institutions)
2. $S1A\_W03$—graduate person has knowledge about relation between social structures
3. $S1A\_W05$—graduate person has knowledge about human-being
4. $S1A\_W06$—graduate person has knowledge about tools and methods used in social analysis
5. $S1A\_W07$—graduate person has knowledge about norms and rules in society
6. $S1A\_U01$—graduate person can interpret social processes
7. $S1A\_U02$—graduate person can use gained knowledge
8. $S1A\_U03$—graduate person can find out reasons

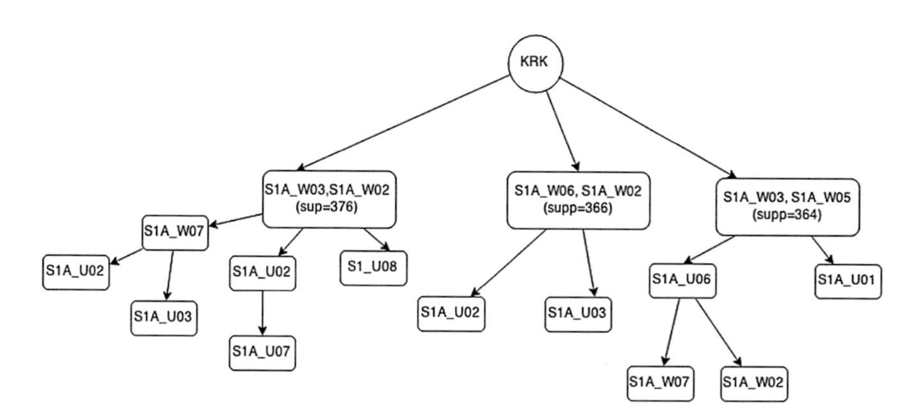

Fig. 3 KRK-based sample tree clusters of patterns

9. $S1A\_U07$—graduate person can analyse the proposed solutions
10. $S1A\_U08$—graduate person can understand social phenomenon

6 Conclusions

We have presented a few approaches to clustering study fields described by the corpora of semi-structured files (types: pdf, doc(x)). First of all we performed information extraction in order to retrieve the KRK competence symbols from files. Next we proposed the various types of clustering methods (flat, hierarchical agglomerative, or pattern based), which goal was to identify equal or similar study fields according to KRK competence vectors. Flat clustering was applied in order to find mirrors clusters (cluster of study fields with equal KRK vectors), which is performed in order to find obvious anomalies in the data. HAC clustering enables us to identify study fields, which educate similar qualifications (the cut-off $= 0.7$ was set relatively high to retain the consistency in the clusters). The last clustering method was the SnSRC method, which finds patterns and organizes them into the tree structures. Such approach gives us possibility to discover regularities in the KRK competence vectors, and present them into the tree structure (with the main component and its sub components). In this step we exploit a novel WSI knowledge-poor algorithm SnS, based on text mining approaches, namely closed frequent sets. Using significant KRK competence patterns SnS builds hierarchical structures called sense frames. Finally, study fields are mapped to the sense frames and clustered accordingly. Additionally, we have built the search engine, which enables to find similar study fields to the given one according to KRK competence similarity. All of those deliverables (search engine, clustering reports) were provided to the Ministry of Science and Higher Education, and were extensively used in order to detect improper KRK competences or discover some knowledge about study fields (educating similar qualifications) and clusters of KRK competence patterns (regularities within KRK competence vectors).

References

1. Andruszkiewicz, P., Nachyla, B.: Automatic extraction of profiles from web pages. In: Intelligent Tools for Building a Scientific Information Platform, Studies in Computational Intelligence. Warsaw (2013)
2. Bernardini, A., Carpineto, C., DAmico, M.: Full-subtopic retrieval with keyphrasebased search results clustering. In: Proceedings of 2009 IEEE/WIC/ACM International Conference on Web Intelligence, pp. 206–213. Milan (2009)
3. Carpineto, C., Osinski, S., Romano, G., Weiss, D.: A survey of web clustering engines. ACM Computing Surveys 41(3), pp. 1–38. ACM, New York (2009)
4. Cutting, D., Karger, D., Pedersen, J., Tukey, J.: Scatter/gather: a cluster based approach to browsing large document collections. In: Proceedings of SIGIR, pp. 318–329. Copenhagen (1992)

5. Di Marco, A., Navigli, R.: Clustering web search results with maximum spanning trees. In: Proceedings of the 12th Congress of the Italian Association for Artificial Intelligence, pp. 201–212. Palermo (2011)
6. Di Marco, A., Navigli, R.: Clustering and diversifying web search results with graph-based word sense induction. In: Computational Linguistics, pp. 709–754. MIT Press, Cambridge (2013)
7. Kozlowski, M.: Word sense discovery using frequent termsets. Ph.D. in Warsaw University of Technology (2014)
8. Kozlowski, M.: Web search results clustering using frequent termset mining. In: Proceedings of 6th International Conference on Pattern Recognition and Machine Intelligence. Warsaw (2015)
9. Kozlowski, M., Rybinski, H.: SnS: A novel word sense induction method. In: Rough Sets and Intelligent Systems Paradigms, pp. 258–268. Madrid (2014)
10. Maarek, I., Fagin, R., Pelleg, D.: Ephemeral document clustering for web applications. IBM Research Report RJ 10186 (2000)
11. Manning, Ch., Raghavan, P., Schutze, H.: Introduction to Information Retrieval. Cambridge University Press, Cambridge (2008)
12. Navigli, R., Crisafulli, G.: Inducing word senses to improve web search result clustering. In: Proceedings of the 2010 Conference on Empirical Methods in Natural Language Processing, pp. 116–126. Boston (2010)
13. Osinski, S., Stefanowski, J., Weiss, D.: Lingo: Search results clustering algorithm based on singular value decomposition. In: Proceedings of the International IIS: IIPWM04 Conference held in Zakopane, pp. 359–368. Zakopane (2004)
14. Van Rijsbergen, C.: Information Retrieval. Butterworths, London (1979)
15. Zaki, M., Hsiao, Ch.: Charm: An efficient algorithm for closed itemset mining. In: Proceedings 2002 SIAM International Conference on Data Mining, pp. 457–472. Arlington (2002)
16. Zamir, O., Etzioni, O.: Web document clustering: A feasibility demonstration. In: Proceedings of the 21st Annual International ACM SIGIR Conference on Research and Development in Information Retrieval, pp. 46–54. New York (1998)
17. Zhang, X., Hu, X., Zhou, X.: A comparative evaluation of different link types on enhancing document clustering. In: Proceedings of SIGIR, pp. 555–562. Singapore (2008)

Semantic Textual Similarity Using Various Approaches

Maciej Kazuła and Marek Kozłowski

Abstract The paper is devoted to the semantic textual similarity (STS) problem. Given two sentences of text, s1 and s2, the systems participating in this problem should compute how similar s1 and s2 are, returning a similarity score. We present our experience in this topic, ranging from the knowledge-poor approaches to some compact and easy applied knowledge-rich methods (using structured knowledge base frameworks like WordNet, Wikipedia or BabelNet). The evaluation of the proposed methods was performed using the datasets from SemEval-2014/15 tasks.

Keywords Semantic textual similarity · Text mining

1 Introduction

Semantic similarity is a crucial component of NLP applications such as Text Summarization, Question Answering, Information Retrieval or Machine Translation. Measure of text similarity was introduced as a vectorial space model in information retrieval [15]. There is a broad scope of other semantic similarity methods based on knowledge-based or corpus-based approaches.

The tasks according to Semantic Textual Similarity appeared in SemEval-2014/15 and aims at an evaluation for semantic similarity across comparable or different sizes of text. The SemEval 2015 task on textual similarity is focused on comparing similar-sized texts. The SemEval 2014 task evaluates the case where larger text must be compared to smaller text, namely there are covered four semantic similarity comparisons: paragraph to sentence, sentence to phrase, phrase to word and word to sense.

M. Kazuła (✉) · M. Kozłowski
National Information Processing Institute, Warsaw, Poland
e-mail: maciej.kazula@opi.org.pl
URL: http://www.opi.org.pl/en/

D. Ryżko et al. (eds.), *Machine Intelligence and Big Data in Industry*,
Studies in Big Data 19, DOI 10.1007/978-3-319-30315-4_5

We present the method for semantic text similarity computation based on a corpus-based or knowledge-based measures between seperated words. We start from preprocessing texts, identifying boundary values, computing terms similarities and derive from them the final score, which is normalized.

The input of the task consists of two text segments of text. We want to determine a score indicating their semantic similarity. Similarity is scored from 0 to 4/5 (depends on the SemEval task 2014 or 2015), when 0 means no semantic intersection, 4/5 means that two items have very similar meanings.

2 Related Work

Text similarity computation has been applied for a long time in applications in Natural Language Processing. One of the earliest applications is the vector space model in Information Retrieval, where the document most relevant to the search query is retrieved and presented in a ranking list, such similarity is called document-level similarity [15, 16].

The typical methods of finding the similarity between two text segments is to use a number of lexical units that occur in both input segments. This approach is successful to a certain degree, this lexical matching similarity fails to identify the semantic similarity of texts using different wordings. For instance, they are not able to detect similarities between texts 'apple is a fruit' and 'tree with flowers'. A significant effort has been made on measuring similarity at the word level, namely by approaches that use distributional semantics like LSA and similar ones [17].

Related work can be classified into four major categories: vector-based document models methods, corpus-based methods, knowledge-based methods and hybrid methods [7].

Vector-based document models represent text as a vector of words and the similarity evaluation is based on the number of words that occur in both texts. Lexical similarity methods have problems with different words sharing common sense. Next approaches, such as corpus-based and knowledge-based methods, overcome the above issues.

Corpus based methods apply scores provided by Pointwise Mutual Information (PMI) and Latent Semantic Analysis (LSA).

The Pointwise Mutual Information (PMI) [18] between two words w_i and w_j is:

$$PMI(w_i, w_j) = log_2 \frac{p(w_i, w_j)}{p(w_i)p(w_j)}$$

The Latent Semantic Analysis (LSA) [8, 9] is a mathematical method for modeling of the meaning of words and contexts by analysis of representative corpora. It models the meaning of words and contexts by projecting them on a vector space of reduced dimensionality, which is built up by applying singular value decomposition (SVD).

Knowledge based methods apply information from semantic networks as WordNet or BabelNet. They exploit the structure of network to compare concepts. There is lots of knowledge-based metrics that were previously proposed based mainly on distance-oriented measures computed on semantic networks. Leacock and Chodorow [10] proposed metric based on the length of the shortest path between two concepts using node-counting. Lesk [11] defined similarity between concepts as the intersection between the corresponding glosses, it was proposed as a solution for word sense disambiguation. Budanitsky and Hirst [5] conducted the research on various WordNet-based measures. Standard thesaurus-based measures of word pair similarity are based only on a single path between concepts. By contrast [6] used a semantic representation of texts from random walks on WordNet. Note that all these metrics are defined between concepts, rather than words, but they can be easily turned into a word-to-word similarity metric by selecting for any given pair of words those two meanings that lead to the highest concept-to-concept similarity.

Hybrid methods use both corpus-based measures and knowledge-based measures of word semantic similarity to determine the text similarity [7]. Mihalcea et al. [13] and Li et al. [12] suggested a combined method by exploiting corpus based measures and knowledge-based measures of words semantic similarity.

The previously mentioned methods are working at document or word level granularity. Pilehvar et al. [14] proposed a generic approach that operates at multiple levels. The method builds a common probabilistic representation over word senses in order to compare different types of linguistic data. Any lexical item is represented as a distribution over a set of word senses (obtained from WordNet), named as item's semantic signature.

3 Approach

Semantic Textual Similarity between two texts is decomposed into similarity between words, which are optionally disambiguated and aligned previously. This means that semantic word-to-word similarity is computed and combined to obtain the text-to-text similarity.

Bag of words has been built as follows—text has been split on delimiters (white spaces and punctuation characters), then stopwords has been removed and remaining words has been lemmatized. Next, two bag of words that has been extracted from texts are noted as sets K and W, where the size of K is smaller than the W one.

Semantic similarity approaches proposed in this paper are knowledge-poor or knowledge-rich. The first one exploits unstructured text corpora without any semantic relatedness between components in order to estimate co-occurrences. The second approach explores knowledge bases like BabelNet or WordNet to create mapping between words and concepts.

In the next sections we describe knowledge-poor and knowledge-rich methods, respectively.

3.1 Knowledge-Poor Methods

The proposed in this section methods are knowledge-poor, it means that they are not using any kind of structured information from resources like machine-readable dictionaries, thesaurus, or ontologies. In addition they are fully unsupervised (there is no learning phase leading to models enable to categorize compared texts). The method exploits only Wikipedia as a raw corpora in order to estimate frequencies of co-occurrences. In order to improve the access to the raw corpora, the Lucene Index has been build for all the documents in the corpora. As long as we need to collect results quickly and we do not have any knowledge resources this approach may be relevant. We were aimed at verifying how good results can be achieved using only corpus-based approach and not including algorithms that have embedded deep language knowledge. Next subsections provide description of two knowledge-poor sub-methods. The key aspects are: to define the word-to-word similarity measure and to combine that partial measure into the final text-to-text similarity score.

Word-to-Word Similarity Measure The used word-to-word similarity measure is the edge case of termset-to-termset measure proposed in [3]. Having two words to compare, we need to query our corpora the number of documents containing words separately, and query the number of documents including both words simultaneously. We run those queries using the Lucene index built upon the corpora. For a given word we define it's cardinality as the number of corpora documents containing the mentioned word. The cardinality of word's conjunction means number of documents where both words appear. The measures are presented below:

$$t2tsim(w,k) = \frac{|w \cap k|}{min(|w|, |k|)} \tag{1}$$

Instead of using minimum function one can take geometric mean, so formula looks then:

$$t2tsim = \frac{|w \cap k|}{\sqrt{|w| \cdot |k|}} \tag{2}$$

The next section described three text-to-text measures consuming the above introduced word-to-word measures.

Text-to-Text Similarity—Local Similarity Oriented Words Alignment The proposed algorithm (as well as the next algorithms) are presented in iterative form. At the beginning we iterate over words from the smaller bag of words K to compute the word-to-word measures against W and find for each iteration the maximal value, which points out the alignment pair of words. In more formal style we can write that procedure in pseudocode as shown in listing of Algorithm 1.

Algorithm 1 Calculate similarity for bag-of-words K and W using local similarity oriented words alignment

---

Require: $|K| \leq |W|$
  $res \leftarrow 0$
  S_k, S_w /* two empty maps */
  for all $k_i \in K$ **do**
    S_k.put(k_i,$|k_i|$) /* $|k_i|$—cardinality of word k_i is the number of documents containing word k_i */
  end for
  for all $w_j \in W$ **do**
    S_w.put(w_j,$|w_j|$) /* $|w_j|$—cardinality of word w_i is the number of documents containing word w_j */
  end for
  for all $k_i \in K$ **do**
    $max \leftarrow -1$
    for all $w_j \in W$ **do**
      retrieve/count from Wiki $|w_j \cap k_i|$ /* cardinality of word's conjunction is the number of documents containing both word k_i and w_j */
      $|k_i| \leftarrow S_k.get(k_i)$
      $|w_j| \leftarrow S_w.get(w_j)$
      $ts \leftarrow \frac{|w_j \cap k_i|}{min(|k_i|,|w_j|)}$ /* calculate partial word-to-word measure */
      if $max < ts$ **then**
        $max \leftarrow ts$
      end if
    end for
    $res \leftarrow res + max$ /* combine partial measures */
  end for
  return res

Text-to-Text Similarity—A Greedy Strategy In the next step, greedy strategy has been applied to align most similar words basing on the all word-to-word measures stored in $n \times m$ two-dimensional table. In each step we find maximal word-to-word similarity and lock the corresponding words as aligned ones. This can be easily achieved by sorting partial results by value in decreasing order and applying greedy strategy. The pseudocode is presented in listing of Algorithm 2.

Algorithm 2 Calculate similarity of bag-of-words K and W using a greedy strategy

Require: $|K| \leq |W|$
  S_k, S_w /* two empty maps */
  $tab[1..i][1..j]$ /* two-dimensional array of word-to-word similarities */
  for all $k_i \in K$ **do**
    S_k.put(k_i,$|k_i|$) /* cardinality of word k_i is the number of documents containing word k_i */
  end for
  for all $w_j \in W$ **do**
    S_w.put(w_j,$|w_j|$) /* cardinality of word w_j is the number of documents containing word w_j */
  end for
  for all $k_i \in K$ **do**
    for all $w_j \in W$ **do**
      retrieve/count $|w_j \cap k_i|$ /* cardinality of word's conjunction is the number of documents containing both word k_i and w_j */
      $|k_i| \leftarrow S_k.get(k_i)$
      $|w_j| \leftarrow S_w.get(w_j)$
      $ts \leftarrow \frac{|w_j \cap k_i|}{min(|k_i|,|w_j|)}$ /* calculate word-to-word measure */
      $tab[i][j] \leftarrow ts$
    end for
  end for
  return SUM- MAXIMALS(tab)

  function SUM- MAXIMALS($tab[1..n][1..m]$))
    $table[n \times m]$ /* one-dimensional table containing all values form the two-dimensional $tab[1..n][1..m]$ */
    S_i, S_j /* two empty sets for word indexes from n and m dimensions respectively */
    $res \leftarrow 0$
    $table \leftarrow tab$ /* store two-dimensional table into one-dimensional table, where values are word-to-word measures */
    $sort(table)$ /* sort table in decreasing order */
    for all $t_k \in table$ **do**
      if $\neg(index\ i$ of $t_k \in S_i\ or\ index\ j\ of\ t_k \in S_j)$ **then**
        $res \leftarrow res + t_k$
        $S_i \leftarrow i$
        $S_j \leftarrow j$
      end if
      if $S_i.size() = n$ or $S_j.size() = m$ **then**
        $break;$
      end if
    end for
    return res
  end function

Text-to-Text—Geometric Mean The last variation of knowledge-poor method is a strategy that uses word-to-word measure based on geometric mean (cosine like measure [1]). In previous algorithms we used min function in denominator, currently we use different expression. We replaced the similarity Eq. 1 with the Eq. 2. Algorithm pseudocode is presented in listing 3.

Algorithm 3 Calculate similarity for bag-of-words K and W based on geometric mean

Require: $|K| \leq |W|$
  $res \leftarrow 0$
  S_k, S_w /* two empty maps */
  for all $k_i \in K$ **do**
    S_k.put(k_i,$|k_i|$) /* cardinality of word k_i is the number of documents containing word k_i */
  end for
  for all $w_j \in W$ **do**
    S_w.put(w_j,$|w_j|$) /* cardinality of word w_j is the number of documents containing word w_j */
  end for
  for all $k_i \in K$ **do**
    $max \leftarrow -1$
    for all $w_j \in W$ **do**
      retrieve/count from Wiki $|w_j \cap k_i|$ /* cardinality of word's conjunction is the number of documents containing both word k_i and w_j */
      $|k_i| \leftarrow S_k.get(k_i)$
      $|w_j| \leftarrow S_w.get(w_j)$
      $ts \leftarrow \frac{|w_j \cap k_i|}{\sqrt{|k_i| \cdot |w_j|}}$ /* calculate word-to-word measures */
      if $max < ts$ **then**
        $max \leftarrow ts$
      end if
    end for
    $res \leftarrow res + max$
  end for
  return res

3.2 *Knowledge-Rich Methods*

Previously, having two arbitrarily ordered texts, we seek the alignment that maximizes the similarity of the words based on relative co-occurrences in the given raw corpora (as Wikipedia). Sometimes such alignment is not enough and there is a strong drive to perform alignment between concepts described by the words. When one requires additional semantic information about entities such method is called knowledge-rich. As a knowledge resource we have used BabelNet. BabelNet is both a multilingual encyclopedic dictionary, with lexicographic and encyclopedic coverage of terms, and a semantic network, which connects concepts and named entities in a very large network of semantic relations, made up of more than 13 million entries, called Babel synsets.

It was decided that BabelNet can be used because it is easily applicable into Java application through well defined API. In case of algorithms exploiting knowledge-rich methods the cardinality of set and conjunction of sets symbols have traditional mathematical meaning.

Text-to-Text—SynSets Oriented Continuing idea of decomposing text-to-text similarity into word-to-word relation, the method of this subsection may be described as follows: for a given word we retrieve SynSets (in practice set of Babelnet concept ids) from BabelNet. Next, for two given words k and w we define $\{k\}$ and $\{w\}$ as sets of corresponding SynSets. The word-to-word measure is defined as

$$t2tsim(k, w) = \frac{|\{k\} \cap \{w\}|}{max(|\{k\}|, |\{w\}|)} \tag{3}$$

Given two SynSets $\{k\}$ and $\{w\}$ for words k and w we divide cardinality of conjunction of those SynSets by maximal cardinality of $|\{k\}|$ and $|\{w\}|$. According to the previous methods there is one significant change—we mutually decrease the SynSets being compared by it's common part. It reflects the idea that some common semantic information in texts is combined into result only once.

Algorithm 4 Calculate similarity for bag-of-words K and W based on SynSets

Require: $|K| \leq |W|$
  $res \leftarrow 0$
  S_k, S_w /* two empty maps */
  for all $k_i \in K$ **do**
    S_k.put(k_i,$\{k_i\}$) /* store SynSets for each word from K in first map */
  end for
  for all $w_j \in W$ **do**
    S_w.put(w_j,$\{w_j\}$) /* store SynSets for each word from W in second map */
  end for
  for all $k_i \in K$ **do**
    $max \leftarrow -1$
    for all $w_j \in W$ **do**
      $\{k_i\} \leftarrow S_k.get(k_i)$
      $\{w_j\} \leftarrow S_w.get(w_j)$
      $ts \leftarrow \frac{|\{w_j\} \cap \{k_i\}|}{max(|\{k_i\}|, |\{w_j\}|)}$ /* calculate word-to-word similarity measure */
      $\{k_i\} \leftarrow \{k_i\} \backslash (\{k_i\} \cap \{w_j\})$
      S_k.put(k_i,$\{k_i|\}$)
      $\{w_j\}| \leftarrow \{w_j\} \backslash (\{w_j\} \cap \{k_i\})$
      S_w.put(w_j,$\{w_j\}$)
      if $max < ts$ **then**
        $max \leftarrow ts$
      end if
    end for
    $res \leftarrow res + max$
  end for
  return res

We verified also the geometric mean in the dominator instead of maximum operator. Modified algorithm is presented in listing 5.

Algorithm 5 Calculate similarity for bag-of-words K and W based on SynSets

Require: $|K| \leq |W|$
  $res \leftarrow 0$
  S_k, S_w /* two empty maps */
  for all $k_i \in K$ **do**
    S_k.put(k_i,$\{k_i\}$) /* store SynSets for each word from K in first map */
  end for
  for all $w_j \in W$ **do**
    S_w.put(w_j,$\{w_j\}$) /* store SynSets for each word from W in second map */
  end for
  for all $k_i \in K$ **do**
    $max \leftarrow -1$
    for all $w_j \in W$ **do**
      $\{k_i\} \leftarrow S_k.get(k_i)$
      $\{w_j\} \leftarrow S_w.get(w_j)$
      $ts \leftarrow \frac{|\{w_j\} \cap \{k_i\}|}{\sqrt{|\{k_i\}| \cdot |\{w_j\}|}}$ /* calculate word-to-word similarity measure * /
      if $max < ts$ **then**
        $max \leftarrow ts$
      end if
    end for
    $res \leftarrow res + max$
  end for
  return res

Text-to-Text—Knowledge-Graph Oriented One of interesting BabelNet features is that it offers some precomputed semantic scores between words which can be used for computing text-to-text similarity. More details are placed in [4]. In our survey we decided to use Knowledge Graph API, which enables to disambiguate words using well-defined BabelNet concepts. The Knowledge Graph API process collection words finding most relevant meanings for each of them (taking into account context information). For the found meaning we retrieve SynSets and further computations are similar to the above described.

Algorithm 6 Calculate similarity for bag-of-words K and W based on Knowledge Graph

Require: $|K| \leq |W|$
  $res \leftarrow 0$
  S_k, S_w /* two empty maps */
  for all $k_i \in K$ **do**
    $k_i \leftarrow meaning\ from\ knowledge\ graph$
    S_k.put(k_i,{k_i}) /* store SynSets for each disambiguated word from K*/
  end for
  for all $w_j \in W$ **do**
    $w_j \leftarrow meaning\ from\ knowledge\ graph$
    S_w.put(w_j,{w_j}) /* store SynSets for each disambiguated word from W */
  end for
  for all $k_i \in keys\ from\ S_k$ **do**
    $max \leftarrow -1$
    for all $w_j \in keys\ from\ S_w$ **do**
      $\{k_i\} \leftarrow S_k.get(k_i)$
      $\{w_j\} \leftarrow S_w.get(w_j)$
      $ts \leftarrow \frac{|\{w_j\} \cap \{k_i\}|}{max(|\{k_i\}|, |\{w_j\}|)}$ /* calculate word-to-word similarity measures */
      if $max < ts$ **then**
        $max \leftarrow ts$
      end if
    end for
    $res \leftarrow res + max$
  end for
  return res

Algorithm 7 Calculate similarity for bag-of-words K and W based on Knowledge Graph

Require: $|K| \leq |W|$
  $res \leftarrow 0$
  S_k, S_w /* two empty maps */
  for all $k_i \in K$ **do**
    $k_i \leftarrow meaning\ from\ knowledge\ graph$
    S_k.put(k_i,{k_i}) /* store SynSets for each disambiguated word from K */
  end for
  for all $w_j \in W$ **do**
    $w_j \leftarrow meaning\ from\ knowledge\ graph$
    S_w.put(w_j,{w_j}) /* store Synsets for each disambiguated word from W */
  end for
  for all $k_i \in keys\ from\ S_k$ **do**
    $max \leftarrow -1$
    for all $w_j \in keys\ from\ S_w$ **do**
      $\{k_i\} \leftarrow S_k.get(k_i)$
      $\{w_j\} \leftarrow S_w.get(w_j)$
      $ts \leftarrow \frac{|\{w_j\} \cap \{k_i\}|}{\sqrt{|\{k_i\}| \cdot |\{w_j\}|}}$ /* calculate word-to-word similarity measure */
      if $max < ts$ **then**
        $max \leftarrow ts$
      end if
    end for
    $res \leftarrow res + max$
  end for
  return res

4 Results

The proposed methods were evaluated within few comparison types using different data sets. The system outputs and the gold standard ratings are compared in two ways, using Pearson correlation and Spearman's rank correlation (rho). Pearson correlation tests the degree of similarity between the system's similarity ratings and the gold standard ratings. Spearman's rho tests the degree of similarity between the rankings of the items according to similarity.

Evaluation was conducted on SemEval 2014 Cross-Level-Seman tic Similarity task3 data sets and SemEval 2015 Semantic Textual Similarity task2 (English subtask) data sets. The appropriate evaluators were provided by SemEval organizers to measure and collect quality metrics. Tables 1, 2 and 3 show results for ours knowledge-poor methods. Tables 4, 5 and 6 contain results for knowledge-rich methods (exploiting the BabelNet structures).

Rules of result's decoration are as follows: results in bold are the best scores among ours methods in a given category/data set (within a table row), results in italics are over SemEval's median score in a given category/data set (within a table row), underlined results are the worst scores among ours methods in a given category/data set (within a table row).

Table 1 Results for SemEval 2015 task-semantic textual similarity—STS-english subtask data sets—Pearson's correlation

Data set	Measure					
	Alg 1	Alg 2	Alg 3	Best	Worst	Median
STS.input.images.txt	0.68816	0.55384	**0.68979**	0.8713	0.0663	0.8036
STS.input.headlines.txt	**0.72472**	0.64645	0.72263	0.8417	–0.0395	0.7418
STS.input.belief.txt	***0.70004***	0.64149	*0.69926*	0.7717	–0.1319	0.6772
STS.input.answers-students.txt	***0.70005***	0.61988	*0.69497*	0.7879	–0.0534	0.6757
STS.input.answers-forums.txt	*0.62551*	0.51303	***0.62561***	0.739	0.1533	0.6132

Table 2 Results for SemEval 2014 task 3 data sets—cross-level semantic similarity—Pearson's correlation

Data set (cross-level similarity type)	Measure					
	Alg 1	Alg 2	Alg 3	Best	Worst	Median
sentence2phrase	0.582174	0.502061	**0.582206**	0.777	0.426	0.68
paragraph2sentence	**0.740201**	0.70969	0.739617	0.837	0.455	0.769

Table 3 Results for SemEval 2014 task 3 data sets-cross-level semantic similarity—Spearman's rho result's

Data set (cross-level similarity type)	Measure					
	Alg 1	Alg 2	Alg 3	Best	Worst	Median
sentence2phrase	0.613107	0.541738	**0.613606**	0.76	0.424	0.674
paragraph2sentence	**0.725923**	0.705207	**0.725923**	0.821	0.533	0.769

Table 4 Results for SemEval 2015 task 2 data sets—semantic textual similarity—STS-english subtask data sets—Pearson's correlation

Data set	Measure						
	Alg 7	Alg 6	Alg 5	Alg 4	Best	Worst	Median
STS.input.images.txt	**0.73752**	0.73749	0.73249	0.73177	0.8713	0.0663	0.8036
STS.input.headlines.txt	0.59943	0.5982	**0.67948**	0.67645	0.8417	−0.0395	0.7418
STS.input.belief.txt	0.59555	0.59466	0.65	**0.65821**	0.7717	−0.1319	0.6772
STS.input.answers-students.txt	0.57723	0.56832	***0.71363***	*0.71357*	0.7879	−0.05344	0.6757
STS.input.answers-forums.txt	0.57626	0.5718	*0.61924*	***0.62446***	0.739	0.1533	0.6132

Table 5 Results form SemEval 2014 task 3 data sets—cross-level semantic similarity—Pearson's correlation

Data set (cross-level similarity type)	Measure						
	Alg 7	Alg 6	Alg 5	Alg 4	Best	Worst	Median
sentence2phrase	0.569988	0.566879	**0.657395**	0.635433	0.777	0.426	0.68
paragraph2sentence	0.720250	0.718710	0.683679	**0.723101**	0.837	0.455	0.769

Table 6 Results for SemEval 2014 task 3 data sets-cross-level semantic similarity—Spearman's rho

Data set (cross-level similarity type)	Measure						
	Alg 7	Alg 6	Alg 5	Alg 4	Best	Worst	Median
sentence2phrase	0.614319	0.614009	***0.683212***	*0.683078*	0.76	0.424	0.674
paragraph2sentence	0.700914	0.700715	0.703664	**0.721720**	0.821	0.533	0.769

Each table contains three additional columns with the best, median and the worst scores of systems taking part in the SemEval competitions (SemEval 2014 subtask 3 and SemEval 2015 subtask 2).

The highest score during whole testing was reported by the Algorithm 1 during tests on paragraph2sentence data set from SemEval 2014 (Table 2, row 2, column 1). The worst result of all tests was achieved by the Algorithm 2 while testing sentence2phrase data set from SemEval 2014 (Table 2, row 1, column 2). Twelve times the results were over median, which means that proposed algorithms are slightly better in comparison to the vast majority of the SemEval competitors. The average difference between the best and the worst results reported by the proposed algorithms in each category is higher in the Knowledge-poor group of methods (about 8 %) in compare to the Knowledge-rich algorithms (about 5 %).

Knowledge-rich methods (Algorithms 4–7) do not outperform permanently the knowledge-poor ones (Algorithms 1–3). In some categories (e.g. STS.input.images or sentence2phrase) the knowledge-rich methods report higher correlation scores than the knowledge-poor algorithms. However, in the categories as e.g. STS.input.headlines or paragraph2sentence, the knowledge-rich solutions are beaten by the the knowledge-poor ones. It means that exploitation of knowledge bases as BabelNet is not a final solution in the semantic textual similarity field.

Time performance analysis is closely dependant on the programming language, which was used. Our approaches were implemented in Java language. Efficiency of those methods is sufficient according to the casual demands (a few seconds per computing the similarity of two phrases). Although knowledge-rich methods are more resource-intensive because they process more complex structures as BabelNet Knowledge Graph.

5 Conclusions and Further Research

In this paper we proposed seven STS methods divided into two groups (knowledge-poor and knowledge-rich). They were evaluated both within few comparison types using various data sets from the SemEval 2014 and 2015. The system outputs and gold standard ratings were compared in two ways, using Pearson correlation and Spearman's rank correlation. Twelve times the results of ours methods were over the SemEval's median, which means that proposed algorithms are slightly better in comparison to other more sophisticated SemEval competitors. The average difference between the best and the worst results reported by the proposed algorithms in each category is higher in the Knowledge-poor group of methods (about 8 %) in compare to the Knowledge-rich algorithms (about 5 %). Knowledge-rich methods do not outperform permanently the knowledge-poor ones. Each data set has the winners from different class of methods. It means that exploitation of knowledge bases as BabelNet is not a final solution in the semantic textual similarity field.

There is one fundamental question concerning semantic textual similarity algorithms—how universal are they? Are they able to work in cross-level environment? One may notice that results of our algorithms are quite proportional to length of the texts so they are not universal—additionally they require scaling. Due to this fact results for phrase2word and word2sense from SemEval 2014 task 3 were not published because they were below the baseline. It means that proposed methods work properly for the phrases and sentences/paragraphs, but not for comparing single words against longer texts.

The level of quality reached for longer texts seems to be enough in order to use them in the production systems, so they are extensively applied in the systems created in National Information Processing Institute. The aim of this work was to analyse methods (knowledge-poor and knowledge-rich) that are easily applicable for measuring semantic similarity.

References

1. http://en.wikipedia.org/wiki/Cosine_similarity
2. http://lucene.apache.org/
3. Kozłowski, M.: OPI: Semeval-2014 task 3 system description. In: Proceedings of the 8th International Workshop on Semantic Evaluation (SemEval 2014), pp. 454–458. Dublin, Ireland, August 23–24 (2014)
4. Navigli, R., Ponzetto, S.: Multilingual WSD with just a few lines of code: The BabelNet API. In: Proceedings of the 50th Annual Meeting of the Association for Computational Linguistics, pp. 67–72. Jeju, Republic of Korea (2012)
5. Budanitsky, A., Hirst, G.: Evaluating WordNet-based measures of lexical semantic relatedness. Comput. Linguist. **32**(1), 13–47 (2006)
6. Hughes, T., Ramage, D.: Lexical semantic relatedness with random graph walk. In: Proceedings of the Conference on Empirical Methods in Natural Language Processing, pp. 581–589 (2007)
7. Islam, A., Inkpen, D.: Semantic text similarity using corpus-based word similarity and string similarity. ACM Trans. Knowl. Discovery Data **2**(2), 1–25 (2008)
8. Landauer, T., Dumais, S.: A solution to Plato's problem: The latent semantic analysis theory of acquisition, induction, and representation of knowledge. Psychol. Rev. **104**, 211–240 (1997)
9. Landauer, T., McNamara, D., Dennis, S., Kintsch, W.: Handbook of Latent Semantic Analysis. Psychology Press (2007)
10. Leacock, C., Chodorow, M.: Combining local context and WordNet sense similarity for word sense identification. In: WordNet, An Electronic Lexical Database, pp. 265–283 (1998)
11. Lesk, M.: Automatic sense disambiguation using machine readable dictionaries: How to tell a pine cone from an ice cream cone. In: Proceedings of the SIGDOC Conference, pp. 24–26 (1986)
12. Li, Y., McLean, D., Bandar, Z., O'Shea, J., Crockett, K.: Sentence similarity based on semantic nets and corpus statistics. IEEE Trans. Knowl. Data Eng. **18**(8), 1138–1149 (2006)
13. Mihalcea, R., Corley, C., Strapparava, C.: Corpus-based and Knowledge-based measures of text semantic similarity. In: Proceedings of the American Association for Artificial Intelligence, pp. 775–780 (2006)
14. Pilehvar, M., Jurgens, D., Navigli, R.: Align, disambiguate and walk: A unified approach for measuring semantic similarity. In: Proceedings of the 51st Annual Meeting of the Association for Computational Linguistics, pp. 1341–1351 (2013)
15. Salton, G., Lesk, M.: Computer Evaluation of Indexing and Text Processing. Prentice-Hall, Englewood Cliffs, New Jersey (1971)
16. Salton, G., McGill, M.: Alternation. In: Introduction to Modern Information Retrieval, McGraw-Hill, New York (1983)
17. Turney, P., Pantel, P.: From frequency to meaning: Vector space models of semantics. J. Artif. Intell. Res. **37**, 141–188 (2010)
18. Turney, P.: Mining the web for synonyms: PMI-IR versus LSA on TOEFL. In: Proceedings of the Twelfth European Conference on Machine Learning, pp. 491–502 (2001)

Part II
Data Mining

Identification of Diabetes Disease Using Committees of Neural Network-Based Classifiers

Ali Hassan El-Baz, Aboul Ella Hassanien and Gerald Schaefer

Abstract Diabetes mellitus is one of the most serious health challenges in both developing and developed countries. In this paper, we present a design of a classifier committee for the detection of diabetes disease based on the Pima Indian diabetic database from the UCI machine learning repository. The proposed method uses multi-layer perceptron (MLP) and cascade-forward back propagation network (CFBN) predictors as base classifiers. The combined committee is based on varying the parameters related to both the design and the training of the neural network classifiers. Our experimental evaluation confirms that the derived approach provides a robust classification system, and yields classification accuracies of 95.31 and 96.88 % based on using combined MLP and combined CFBN classifiers respectively. The experimental results obtained thus show that the proposed classifier committee can form as useful basis for automatic diagnosis of diabetes.

1 Introduction

Diabetes mellitus is one of the most serious health challenges in both developing and developed countries. According to the International Diabetes Federation (IDF), 382 million people suffered from diabetes in 2013 while this is expected to rise to 592 million by 2035. Diabetes mellitus is a complex metabolic disorder which is characterised by persistent hyperglycemia and results from defects in insulin secretion, insulin action or both [1]. Diabetes can be either Type-1 (insulin dependent) or Type-2 (non-insulin dependent). The dataset we are focussing on in this paper originates from a population of Pima Indians and comprises Type-2 positive and negative

A. Hassan El-Baz (✉)
Department of Mathematics, Damietta University, Damietta, Egypt
e-mail: ali_elbaz@yahoo.com

A. Ella Hassanien
Faculty of Computers and Information, Cairo University, Giza, Egypt

G. Schaefer
Department of Computer Science, Loughborough University, Loughborough, UK
e-mail: gerald.schaefer@ieee.org

D. Ryżko et al. (eds.), *Machine Intelligence and Big Data in Industry*,
Studies in Big Data 19, DOI 10.1007/978-3-319-30315-4_6

instances. According to [2], the Pima Indians of Arizona have the highest reported incidence of diabetes in the world, while the predominant Type-2 diabetes observed there is typically characterised by slow and gradual commencement. Consequently, the traditional diagnostic method which is partially based on the plasma glucose test may be delayed by up to ten years [1].

Biomedical signal classification can be defined as a categorisation of the recorded input data, which is often based on extracting significant features into distinct classes [3]. The objective is to group the observed signals, thus generating decision boundaries to separate the involved categories. Typical examples include diagnostic checks so as to identify whether a patient suffers from a certain disease such as diabetes or not.

While the traditional approach used in designing pattern classification systems is based on experimental comparison of several classification algorithms in order to select the best one [4], more recently combinations of multiple classifier decisions are investigated [4–12]. A combined classifier can perform better than any of its base continuant classifiers due to the complementary performance of the base classifiers which leads to a more stable combination. A diverse set of base classifiers can be generated by parameter variation of the base classifier and/or variations of the training sets as in bagging and boosting approaches [13–15]. Ensembles of neural network classifiers have been introduced in [16–18]. The base classifiers are different from each other in one or more aspect such as the training parameters, initial weights, network architecture or training algorithm.

In this paper, we present an effective approach to diabetes classification based on a committee of neural network classifiers. We enforce diversity in the ensemble by varying the architectures of the neural networks, with the involved parameters chosen adaptively to generate a pool of classifiers. The final decisions are combined using a majority voting rule, and the resulting ensemble is shown to give excellent classification performance on the Pima Indian diabetic database from the UCI machine learning repository.

2 Diabetes Dataset

The dataset used for our study is the Pima Indians Diabetes Database (PIDD) from the UCI machine learning repository.[1] The dataset is a collection of medical diagnostic reports from 768 records of female patients of Pima Indian heritage in Arizona. The following attributes form each record: number of times pregnant, plasma glucose concentration, diastolic blood pressure, triceps skin fold thickness, 2-hour serum insulin, body mass index, diabetes pedigree function, and age. Each subject was tested as either positive or negative with 500 non-diabetic patients and 268 diabetic ones.

[1] Available at http://archive.ics.uci.edu/ml/datasets/Pima+Indians+Diabetes.

3 Proposed Classifier Committee

A variety of classifiers has been used for diagnosis of diabetes including nearest neighbour, multi-layer perceptron, decision tree, Bayes, and fuzzy classification algorithms to name a few. However, rather than relying on a single predictor, combining multiple classifiers can lead to improved and more robust classification performance. In this paper, we design such a combined classifier, or classifier ensemble, which makes use of parameter based diversity enforcement, for the purpose of diabetes diagnosis.

3.1 Multi-layer Perceptron (MLP) Neural Network

Multi-layer neural networks (MLNNs) have been successfully used in replacing conventional pattern recognition methods for a variety of applications including disease diagnosis [19, 20]. In the first stage of this study, a multi-layer neural network structure was used for diabetes identification on the Pima Indian dataset. The network structure consists of input layer, hidden layers, and output layer, where the hidden layer output layer neurons use non-linear sigmoid activation functions. There are eight inputs features, while the two outputs relate to the two classes. We used the Levenberg-Marquardt algorithm [21] for training the neural networks.

3.2 Cascade-Forward Back Propagation Network (CFBN)

A cascade type of feed-forward neural networks consists of a layer of input, a layer of output neurons, and one or more hidden layers. Similar to a general feed-forward networks, the first layer has weights coming from the input. However, each subsequent layer has weights coming from the input and all previous layers, while all layers have biases. The last layer is the network output. It was found [22] that CFBN can provide a more effective algorithm compared to feed-forward back propagation methods.

3.3 Proposed Classifier Ensemble

Combining multiple classifiers has been intensively studied and is widely regarded as an effective technique to overcome the limitations of individual classifiers [31, 32] with many well-known models having been proposed [33, 34]. Importantly, these classifier ensemble methods, such as neural network ensembles, have been applied successfully in many real-world applications [24, 25].

In general, there are two main categories of classifier ensembles. The first one aims at learning multiple classifiers at the feature level, and usually trains and combines multiple classifiers in the learning process, such as boosting [26] and bagging [14]. The second group tries to combine classifiers at the output level, where the results of multiple available classifiers are combined, e.g., multiple classifier systems or mixtures of experts [27]. In this paper, we focus on the second approach. In particular, given multiple classifiers (available or sequently learned, homogeneous or heterogeneous), the classifier ensemble is learned by combining these component classifiers in an intelligent way.

Classifiers differing in feature representation, architecture, learning algorithm or training data exhibit complementary classification behaviour and the fusion of their decisions can yield higher performance than the best individual classifier. It had been observed that although one classifier would yield the best performance, the sets of patterns misclassified by different classifiers would not necessarily overlap. This suggests that different classifiers potentially offer complementary information about the patterns to be classified, which can be exploited in order to improve the overall classification performance. By combining classifiers, the idea is not to rely on a single predictor but instead combine the individual opinions of multiple classifiers to arrive at a consensus decision.

In [28], three approaches to building multiple classifier systems were presented: divide and conquer approach, sequential approach and parallel approach. In the divide and conquer approach, the types of input are isolated on which each specific classifier performs well, and a new input is directed accordingly. In the sequential approach, one classifier is used at the beginning and other classifiers are invoked only if the first classifier fails, while in the parallel approach, all individual classifiers are applied in parallel to the same input data, and a combination method is applied to aggregate their decisions. It is this last technique that we employ in this paper.

In particular, we train—randomly—many individual neural network classifiers. Each classifier is different from the others in terms of number of epochs, learning rate, and/or number of hidden neurons of the network. At the end, we combine the classifier results by using the majority voting technique (also known as plurality voting technique). Both multi-layer neural network and cascade-forward back propagation network are used as base classifier in our proposed ensemble systems.

3.4 Combination Rules

Fixed combiners are widely studied in the literature on combining classifiers [29, 30]. The confidence $q_j(x)$ for class j is computed by

$$q_j(x) = \frac{h_j(x)}{\sum_i h_j(x)} \tag{1}$$

with

$$h_j(x) = rule_i(p_{ij}(x)). \tag{2}$$

The final classification can be made by

$$w(x) = \arg_j \max(q_j(x)). \tag{3}$$

This Maximum rule selects the classifier producing the highest estimated confidence.

In contrast, in ensemble majority voting, which is a popular way of combining classifiers, the votes for each class over the input classifiers are counted and the majority class selected. In terms of the rule of Eq. 2 this equates to

$$h_j(x) = \sum_i I(\arg_i \max(q_j(x)) = i), \tag{4}$$

where $I()$ is the indicator function defined as

$$I(y) = \begin{cases} 1, & \text{if y = true;} \\ 0, & \text{otherwise} \end{cases}. \tag{5}$$

4 Experimental Results

4.1 MLP Ensemble

In the first experiment, MLP is used as base classifier for our proposed ensemble. In order to generate diversity in the classifier ensemble, a pool of 16 different classifiers are generated and the final hypothesis is combined using majority voting. The classification accuracy of the best individual classifier on the test data is 80.21 % while the proposed combined classifier achieved a classification accuracy of 95.31 %. We note that the proposed combined classifier compares favourably with results from the literature. Figure 1 shows the accuracy of base MLP versus the number of individual classifiers, while Figs. 2 and 3 show the accuracy of base MLP versus the number of hidden neurons and number of training epochs, respectively.

4.2 CFBN Ensemble

In the second experiment, a cascade-forward back propagation network (CFBN) is used as the base classifier for the combined classifier. The best individual classifier gave a classification accuracy of 81.77 % on the test data. In contrast, the proposed combined classifier achieved a performance of 96.88 %. The CFBN approach thus gave slightly better compared to the MLP, both for the best individual classifier and

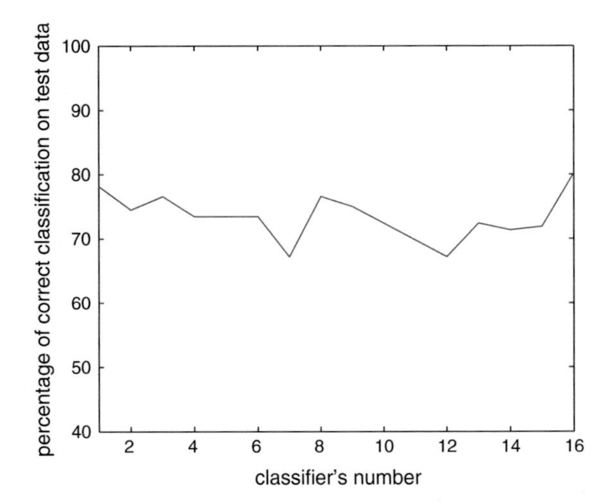

Fig. 1 Accuracy of base MLP versus the number of individual classifiers

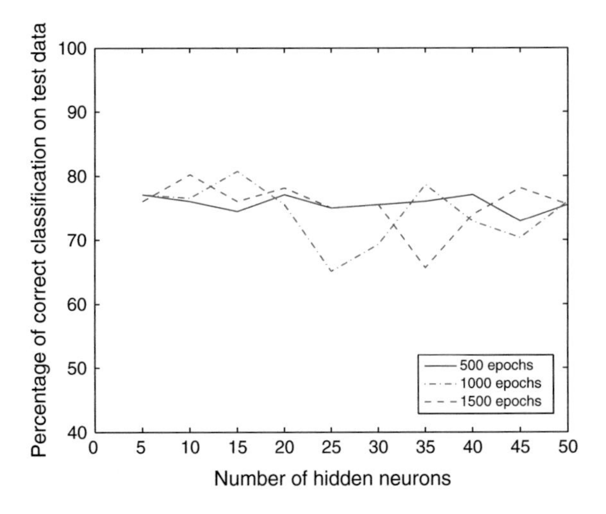

Fig. 2 Accuracy of base MLP versus the number of hidden neurons

for the generated ensemble. Figure 4 shows the accuracy of base CFBN versus the number of individual classifier, while Figs. 5 and 6 show the accuracy of base CFBN versus the number of hidden neurons and number of training epochs, respectively.

5 Performance Measurements

There are a number of metrics to measure the performance of classification methods which are commonly used in medical diagnosis tasks, which are based on the definitions of true positives (TP, the number of correct predictions in which an instance is positive), true negatives (TN, the number of incorrect predictions in which an in-

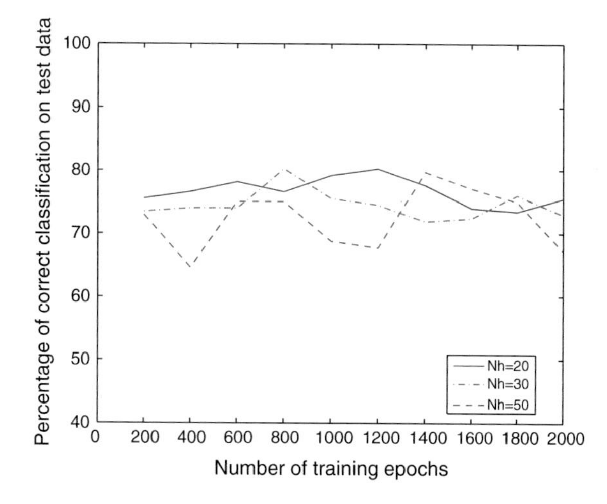

Fig. 3 Accuracy of base MLP versus the number of training epochs

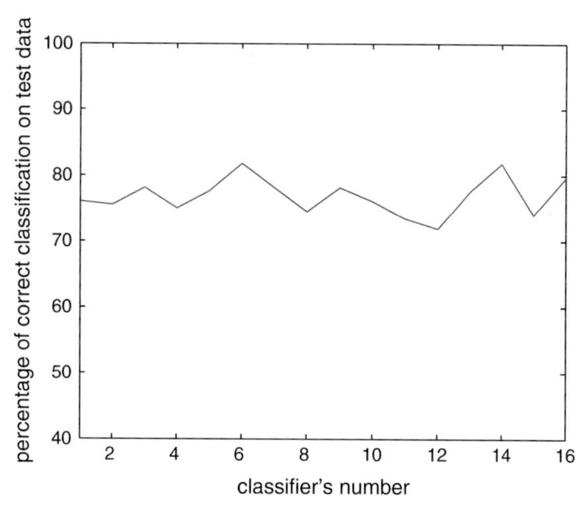

Fig. 4 Accuracy of base CFBN versus the number of individual classifiers

stance is negative), false positives (FP, the number of incorrect predictions in which an instance is positive), and false negatives (FN, the number of correct predictions in which an instance is negative).

Recall (sensitivity) is the percentage of real positive cases that are correctly predicted positive. Conversely, precision (confidence) indicates the percentage of predicted positive cases that are correctly real positives. The F-measure combines precision and recall as their harmonic mean. Specificity indicates the percentage of samples that were classified as normal and which were labeled as normal. In Fig. 7, we summarise the experimental results, in terms of these measures. As we can see, the performance of combining CFBNs outperforms an ensemble of MLPs.

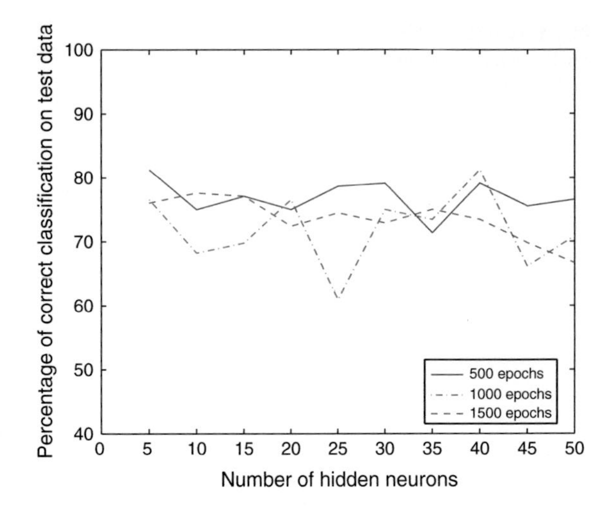

Fig. 5 Accuracy of base CFBN versus the number of hidden neurons

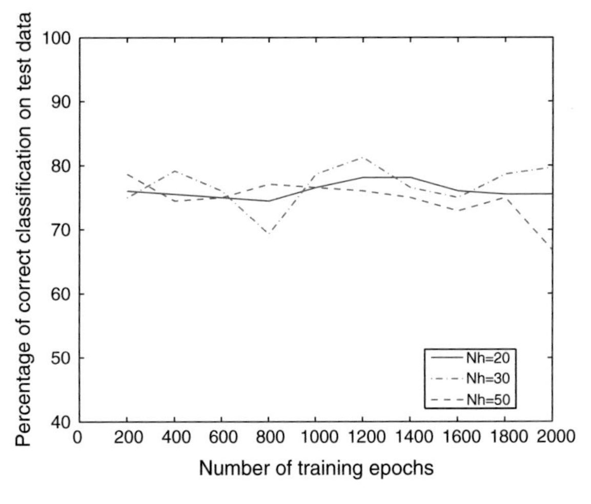

Fig. 6 Accuracy of base CFBN versus the number of training epochs

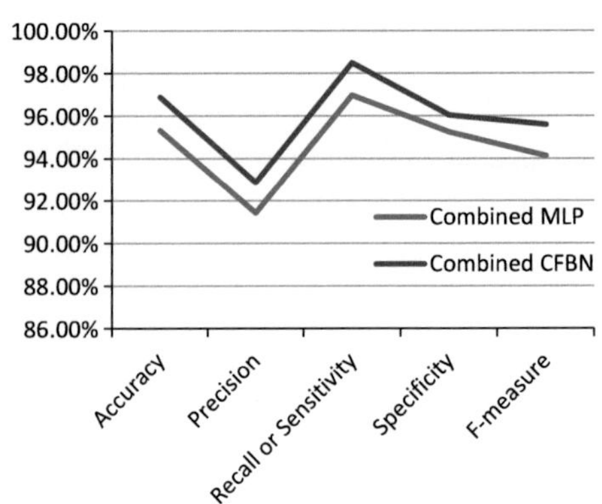

Fig. 7 Performance measures of the proposed methods on test data

6 Conclusions

Automatic identification of diabetes is a challenging task in pattern recognition. Combinations of weak classifiers have shown the potential to achieve good generalisation performance. In this paper, we have introduced a classifier ensemble to solve the instability problem of individual pattern classifiers in the context of diabetes diagnosis. The proposed classifier consists of multiple individual classifiers which are different in both architecture and training parameters. As confirmed in the experimental evaluation, the achieved classification accuracies of the combined classifiers clearly outperform those of the best individual classifiers. The performance is further found to outperform that of the reported methods in the literature. Of the two investigated neural network models that were employed as base classifiers, cascade-forward back propagation networks were found to outperform multi-layer perceptron networks, both as individual classifiers and in terms of the resulting ensembles.

References

1. Holt, R., Hanley, N.: Essential Endocrinology and Diabetes. Blackwell Publishing, Malden (2006)
2. Knowler, W.C., Bennett, P.H., Bottazzo, G.F., Doniach, D.: Islet cell antibodies and diabetes mellitus in Pima Indians. Diabetologia **17**(3), 161–174 (1979). Springer, Heidelberg, Berlin
3. Rangayyan, R.M.: Biomedical Signal Analaysis, A Case-study Approach. IEEE Press series, Canada (2002)
4. Kittler, J., Hatef, M., Duin, R.P.W., Matas, J.: On combining classifiers. IEEE Trans. Pattern Anal. Mach. Intell. **3**, 226–239 (1998)
5. Rogova, G.: Combining the results of several neural network classifiers. Neural Netw. **7**(5), 777–781 (1994)
6. Zhou, Z.H., Wu, J., Tang, W.: Ensembling neural networks: Many could be better than all. Artif. Intel. J. **137**(1–2), 239–263 (2002)
7. Kuncheva, L.I.: Combining Pattern Classifiers: Methods and Algorithms. Wiley-Interscience, New York (2004)
8. Garca-Osorio, Csar, de Haro-Garca, Aida, Garcia-Pedrajas, Nicols: Democratic instance selection: A linear complexity instanceselection algorithm based on classifier ensemble concepts. Artif. Intell. J. **174**(5–6), 410–441 (2010)
9. Galar, M., Fernandez, A., Barrenechea, E., Bustince, H., Herrera, F.: An overview of ensemble methods for binary classifiers in multi-class problems: experimental study on one-vs-one and one-vs-all schemes. Pattern Recogn. **44**(8), 1761–1776 (2011)
10. Tabassian, M., Ghaderi, R., Ebrahimpour, R.: Combination of multiple diverse classifiers using belief functions for handling data with imperfect labels. Expert Syst. Appl. **39**(2), 1698–1707 (2012)
11. Boln-Canedo, V., Snchez-Maroo, N., Alonso-Betanzos, A.: An ensemble of filters and classifiers for microarray data classification. Pattern Recogn. **45**(1), 531–539 (2012)
12. Podolak, I.T., Roman, A.: Cores: Fusion of supervised and unsupervised training methods for a multi-class classification problem. Pattern Anal. Appl. 1–19 (2011)
13. Diettrich, T.G.: Ensemble methods in machine learning. In: Proceedings of the first international workshop on multiple classifier systems. Lecture Notes in Computer Science, pp. 1–15. June, Springer, Berlin (2000) (Cagliari, Italy)
14. Breiman, L.: Bagging predictors. Mach. Learn. **24**(2), 123–140 (1997)

15. Freund, Y., Schapire, R.E.: A decision-theoretic generalization of on-line learning and an application to boosting. J. Comput. Syst. Sci. **55**(1), 119–139 (1997)
16. Giacinto, G., Roli, F.: Design of effective neural network ensembles for image classification. J. Image Vis. Comput. **19**(9–10), 697–705 (2001)
17. Sharkey, A.J.C.: On combining artificial neural nets. Connection Sci. **8**, 299–314 (1996)
18. Sharkey, A.J.C., Sharkey, N.E., Gerecke, U., Chandroth, G.O.: The test and select approach to ensemble combination. In: Kittler, J., Roli, F. (Eds.) Proceedings of the first International Workshop on Multiple Classifier Systems (MCS2000). LNCS 1857, pp 30–44. Springer, New York (2000)
19. Delen, D., Walker, G., Kadam, A.: Predicting breast cancer survivability: A comparison of three data mining methods. Artif. Intell. Med. **34**(2), 113–127 (2005)
20. Polat, K., Gunes, S.: An expert system approach based on principal component analysis and adaptive neuro-fuzzy inference system to diagnosis of diabetes disease. Digital Signal Process. **17**(4), 702–710 (2007)
21. Hagan, M.T., Menhaj, M.: Training feed forward networks with the Marquardt algorithm. IEEE Trans. Neural Netw. **5**, 989–993 (1994)
22. Filik, U.B., Kurban, M.: A new approach for the short-term load forecasting with autoregressive and artificial neural network models. Int. J. Comput. Intell. Res. **3**(1), 66–71 (2007)
23. Sokolova, M., Japkowicz, N., Szpakowicz, S.: Beyond accuracy, F-score and ROC: A family of discriminant measures for performance evaluation. AI 2006: Adv. Artif. Intell. 1015–1021 (2006)
24. Dasarathy, B.V.: A special issue on applications of ensemble methods. Inf. Fus. **9**(1)(2008)
25. Oza, C., Tumer, K.: Applications of ensemble methods. Inf. Fus. **9**(1), 2–3 (2008)
26. Schapire, R.: The strength of weak learnability. Mach. Learn. **5**(2), 197–227 (1990)
27. Yuksel, S.E., Wilson, J.N., Gader, P.D.: Twenty years of mixture of experts. IEEE Trans. Neural Netw. Learn. Syst. **23**(8), 1177–1193 (2012)
28. Ho, T.K.: Data complexity analysis for classifier combination. In: Proceedings of the 2nd International Workshop of Multiple Classifier System 2096. Lecture Notes in Computer Science, pp. 53–67. Springer, Cambridge (2001)
29. Kittler, J., Hatef, M., Duin, R.P.W., Matas, J.: On combining classifiers. IEEE Trans. Pattern Anal. Mach. Intell. **20**(3), 226–239 (1998)
30. Xu, L., Krzyzak, A., Suen, C.Y.: Methods of combining multiple classifiers and their applications to handwriting recognition. IEEE Trans. Systems, Man, Cybern. **22**(3), 418–435 (1992)
31. Ho, T., Hull, J., Srihari, S.: Decision combination in multiple classifier systems. IEEE Trans. Pattern Anal. Mach. Intell. **16**(1), 66–75 (1994)
32. Dietterich, T.: Ensemble method in machine learning. In: Proceedings of International Workshop on Multiple Classifier Systems, pp. 1–15 (2000)
33. Krogh, A., Vedelsby, J.: Neural network ensembles, cross validation, and active learning. In: Advances in Neural Information Processing Systems, vol. 7, pp. 231–238 (1995)
34. Hansen, L.K., Salamon, P.: Neural network ensembles. IEEE Trans. Pattern Anal. Mach. Intell. **12**(10), 993–1001 (1990)

Enzyme Function Classification Based on Borda Count Ranking Aggregation Method

Mahir M. Sharif, Alaa Tharwat, Aboul Ella Hassanien, Hesham A. Hefny and Gerald Schaefer

Abstract Prediction of enzyme functions is an important research topic due to their role in chemical reactions. In this paper, we propose a model for enzyme function classification that combines the outputs of different pairwise sequence alignments based on local sequence alignment. The output of each pairwise sequence alignment is represented by a ranked list, while the main idea of the proposed model is to combine all ranked lists into one ranked list. The candidate of the highest rank is then assigned as the function of the unknown sequence. Unbalanced and balanced datasets are used for evaluation, and the obtained results show that our approach yields good performance and that ranking aggregation achieves results better compared to all single sequence alignments.

Keywords Enzyme classification · Local sequence alignment · Ranking aggregation · Borda count

1 Introduction

The process of enzyme classification and prediction remains a topic of interest in various research areas. Approaches for enzyme prediction and classification are based on protein structure, protein features, and sequence alignment. Classifying

M.M. Sharif (✉) · H.A. Hefny
Institute of Statistical Studies and Researches (ISSR), Cairo University, Giza, Egypt
e-mail: mahiralsharif@yahoo.com

A. Tharwat
Faculty of Engineering, Suez Canal University, Ismaileya, Egypt

A.E. Hassanien
Faculty of Computers and Information, Cairo University, Giza, Egypt

G. Schaefer
Department of Computer Science, Loughborough University, Loughborough, UK
e-mail: gerald.schaefer@ieee.org

D. Ryżko et al. (eds.), *Machine Intelligence and Big Data in Industry*, Studies in Big Data 19, DOI 10.1007/978-3-319-30315-4_7

enzymes based on their structure is not widely used. Of recent approaches using simple sequences, structure properties, and features extracted from the enzymes [2], the latter is based on sequence similarity measures, and this analysis is the most common way of functional prediction, with the majority of sequences in protein databases annotated using just sequence comparison [11, 14].

According to the enzyme commission (EC), enzyme functions are classified into six main classes including:

- Oxidoreductase: catalysing oxidation-reduction reactions; the systematic enzyme name is in the form donor: acceptor oxidoreductase. The substrate that is being oxidised is regarded as the hydrogen donor;
- Transferase: transferring a chemical group from one substrate (the donor) to another (the acceptor) according to a general reaction $XY + Z = X + YZ$;
- Hydrolase: these enzymes catalyse the hydrolytic cleavage of bonds such as C–O, C–N, C–C and some other bonds, including phosphoric anhydride bonds;
- Lyase: catalysing the non-hydrolytic and cleave C–C, C–O, C–N and other bonds by means other than hydrolysis or oxidation;
- Isomerase: catalysing geometrical or structural changes within one molecule. According to the type of isomerism involved they are called racemases, epimerases, cistrans-isomerases, isomerases, tautomerases, mutases or cycloisomerases;
- Ligase: catalysing the joining of two molecules coupled with hydrolysis of a pyrophosphate bond in ATP or a similar triphosphate. The systematic enzyme name takes the form A:B ligase (XDP or XMP-forming). The recommended name often takes the form A–B ligase [7].

There are many studies focussing on enzyme function prediction. Faria et al. [2] developed a machine learning methodology called Peptide Program that achieved high accuracy based on a small dataset. Many other studies were based on extracting features from sequences. Mohammed and Guda [8] reviewed a group of studies, which used a number of computational models to predict and classify enzymes based on extracted features such as molecular weight, polar, etc. Various approaches predict the function of enzymes based on combining two different methods. Xing et al. and Omer et al. presented a method for extracting features from motif content and protein composition for protein sequence classification [9]. Tian and Skolnick [14] used both function and sequence similarities and studied the conservation of enzyme functions by averaging the functional conservation across all enzyme families. Bum et al. proposed new PNPRD features representing global and/or local differences in sequences, based on positively and/or negatively charged residues, to assist in predicting protein function, and used feature subsets for predicting the function of various proteins [6]. Chien et al. reported that some specific features are physiochemical properties of protein complex subunits [4].

Combining results from different independent resources can achieve more accurate results. For example, in biometrics and other identification systems, combining features and independent classifiers can increase the performance and robustness of the system [5, 12].

In this paper, our proposed model aims to calculate the pairwise alignment between the unknown and all training sequences using several local sequence similarity alignments due to its robustness against global alignment [10]. Different outputs of pairwise local alignments are calculated based on different score matrices and represented as a list of ranks. Each ranked list has the class label or enzyme function of all sequences sorted from the nearest one to the farthest. Then, we combine all ranked lists using the Borda count method into one final ranked list. The class label of the highest ranked element in the final ranked list the represents the enzyme function of the unknown sequence.

The rest of the paper is organised as follows. In Sect. 2, the theoretical framework of our proposed model is explained. Section 3 then introduces the steps of our approach in detail. Section 4 presents the experimental results, while Sect. 5 concludes the paper.

2 Preliminaries

In this section, the theoretical framework of the proposed model is explained. First, the sequence alignment task is introduced including how to align two sequences and different types of score matrices. Then, local alignment is briefly discussed. Finally, ranking aggregation techniques are explained, including the Borda count method.

2.1 Sequence Alignment

Sequence alignment is a process of arranging two sequences to determine the region of similarity and similarity score between these two sequences.

Given two sequences X and Y, where $X \equiv x_1 x_2 \dots x_m$, $Y \equiv y_1 y_2 \dots y_n$, and x_i and y_i are letters chosen from the alphabet, the distance or similarity between X and Y represents the number of similar characters between them. In each sequence, gaps of an arbitrary number of null characters or spaces may be placed. The alignment of a character with null is called indel. Indel may be understood as an insertion of a character into one sequence or deletion of a character from the other one [10].

After aligning the pair of sequences, we need to measure the resulting score of alignment or similarity. Calculating the scores depends on a scoring or substitution matrix [10], which is a set of values representing the likelihood of one residue being substituted by another. Two well-known scoring matrices for proteins are Point Accepted Mutation or Percent Accepted Mutation (PAM) and BLOck Substitutions Matrices (BLOSUM) [10, 15].

The alignment score is the sum of the scores for aligning pairs of letters (alignment of two letters) and gap scores (alignment of a gap with letter). Two well-known techniques in pairwise sequence alignment are global and local sequence alignment techniques [10].

2.2 Local Alignment

The first algorithm that used local alignment to measure the similarity between different sequences is proposed by Smith–Waterman as

$$SIM(i,j) = \max \begin{cases} SIM(i-1,j-1) + s(x_i, y_i) & \text{match or mismatch} \\ SIM(i-1,j) + g & \text{insertion gap} \\ SIM(i,j-1) + g & \text{deletion gap} \\ 0 & \text{otherwise} \end{cases} \quad (1)$$

where g represents a gap, and $s(x_i, y_i)$ represents the matching or mismatching scores between x_i and y_i.

Local alignment is useful when the sequences are not similar in length or characters, but it finds the most similar regions in two sequences being aligned [15]. Thus, local alignment isolates regions in the sequences, and hence it is easy to detect repeats.

2.3 Ranking Aggregation

Ranking aggregation methods can be classified into Condorcet and non-Condorcet methods. A Condorcet method [16] is a single-winner election that ranks the candidates in order of preference. The idea is based on pairwise voting; it compares every possible pair of candidates in the ranked lists to decide the preference of them. In contrast, in non-Condorcet methods, the aggregation is calculated based on the location of each candidate in each list. There are various non-Condorcet aggregation methods such as Borda count, highest rank fusion, and logistic regression rank fusion.

In our model, the ranking aggregation method is applied on the ranked lists that are calculated by different local pairwise sequence alignments based on different score matrices. For ranking aggregation, the output of each local alignment process is represented by a ranked list R, where the list represents a set of enzyme functions of the training sequences sorted in decreasing order of confidence. Thus, each class has many ranks and its ranks depend on the distances between the unknown sequence and all other training sequences.

For example, if we have three different outputs (ranked lists) of different sequence alignments R_1, R_2, and R_3, each ranked list is sorted from the highest to the lowest as $R_i =< a, d, c, b, e, f >$, where a, d, c, b, e and f represent the candidates or elements of the ranked list R_i and $f < e < d < c < b < a$. In our model, a, d, c, b, e and f represent the enzyme functions determined from the alignment between the unknown sequence and all the training sequences. The fusion of all ranked lists is used to calculate the final ranked list and determines the decision, i.e. the predicted enzyme function.

2.4 Borda Count Ranking Aggregation

The aim of the Borda count method is to select the class label that has the highest rank of multiple ranked lists. The score of the highest ranked decision is $(n - 1)$ when the number of classes is n and the second highest ranked class gets the score of $(n - 2)$, etc. [3].

As an example, assume the number of ranked lists $L = 4$, and the number of classes or functions $c = 6$. Assume that for a certain unknown sequence x the outputs of each sequence alignment represented by class labels sorted in descending order and the most accurate result is the highest rank as shown in Fig. 1. The outputs of each alignment process are represented by a ranked list. The idea of the Borda count method is to combine or add the ranks of each class and then sort the class labels according to their new fused scores [5, 13]. Thus, in the given example, as shown in Fig. 1, the unknown sequence x is assigned to class $C1$ which aggregates the highest ranks.

The detailed steps of the Borda count method are explained in Algorithm 1.

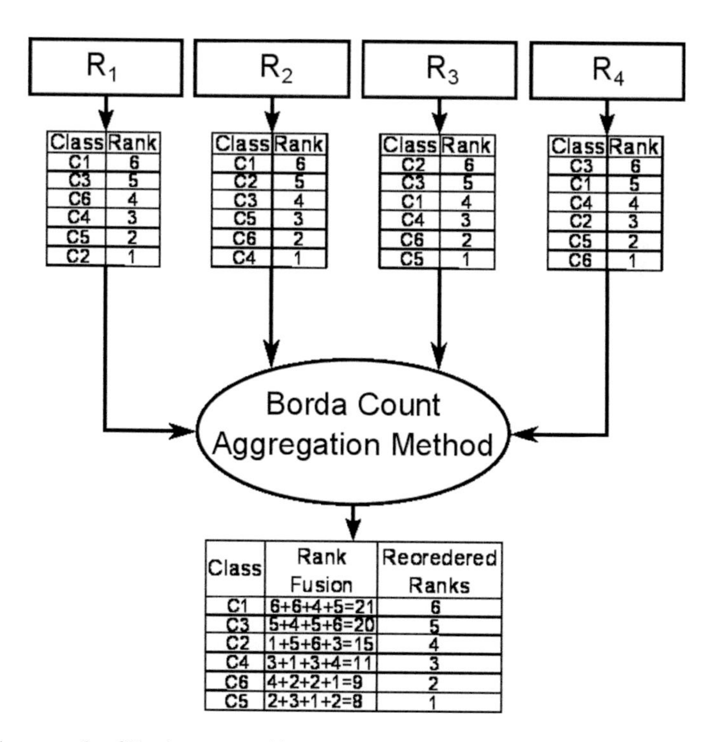

Fig. 1 An example of Borda count ranking aggregation method

Algorithm 1 Borda Count Algorithm

Get L ranking lists from L sequence alignments; each list has N identities sorted in descending order, where the higher order means higher rank.
for all ranking lists (L) **do**
  for all identities (N) **do**
    Calculate the total Borda score of each identity as $B_c = \sum_{i=1}^{L} B_i$, where B_i represents the score or rank of the ith ranking list.
  end for
end for
Sort B_c in descending order and replace with corresponding identity.

---

3 Proposed Enzyme Function Classification Model

Our proposed model is illustrated in Fig. 2 and comprises two main steps. In the first step, we apply different pairwise sequence similarity alignments between the unknown sequence and all the training or labelled sequences. All pairwise alignments are local alignments, and different pairwise alignments are calculated based on different scoring matrices. The outputs from each pairwise alignment are represented by a ranked list. Each list has the enzyme functions ranked from the highest rank, which is the nearest score (highest score) to the unknown sequence, to the lowest rank, which represents the enzyme function of the farthest element (the element that has the minimum score). After calculating all ranked lists, the next step is to combine

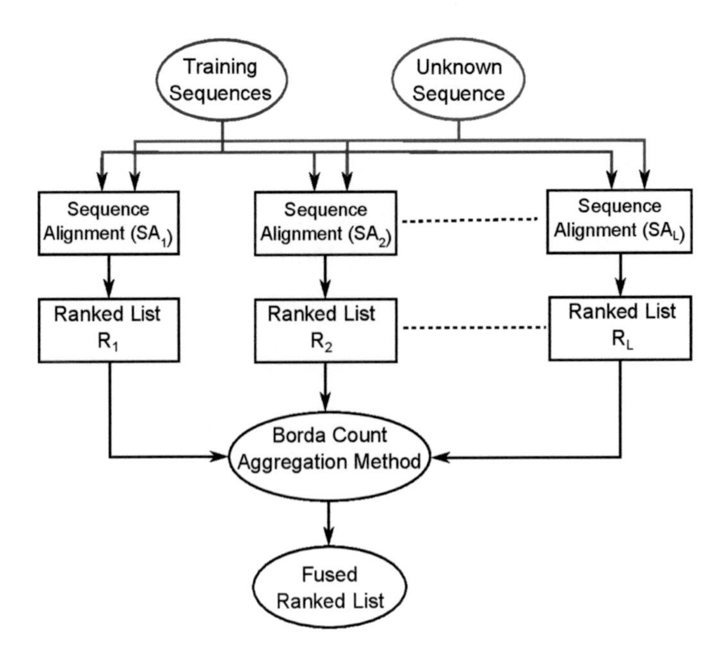

Fig. 2 Block diagram of the proposed model

or fuse the ranked lists using the Borda count method. The results of the Borda count are represented by a new ranked list, where the highest score represents the class label of an enzyme function of the unknown sequence.

The details of the proposed steps are summarised in Algorithm 2.

Algorithm 2 Enzyme function classification Based on the proposed model

---

Input: A set of labelled sequences with their functions $S_i, i = 1, 2, \ldots, N$ where N represents the total number of sequences.

Given an unknown sequence T, align T with all labelled sequences S using local pairwise alignment based on different score matrices.
for all $i = 1, 2, \ldots, L$, where L is the number of ranked lists/alignments **do**
  Compute the outputs of each sequence alignment process as a ranked list R_i
end for
Combine the ranked lists to get one final ranked list R_{final}.
Assign the class label of the highest rank element which represents its enzyme function to the unknown sequence T.

---

4 Experimental Results

For evaluation, we designed two experimental scenarios to test our proposed model. In all cases, we used a one against all approach; in other words, for each unknown sequence, we calculate the sequence alignment (SA) between an unknown sequence and all other sequences in the dataset. In all experiments, performance is evaluated using accuracy (i.e., the percentage of the total number of predictions that were correct).

4.1 Dataset

The dataset that is used in all experiments is obtained from the SWISS-PROT enzyme database which consists of 6923 sequences of enzymes from [1]. The details of the classes are shown in Table 1. We used a subset which represents approximately half ($\frac{3470}{6923} = 50.12\,\%$) of the total number of all sequences % as shown in Table 1.

4.2 Experiment Scenarios

In our first experiment, the unknown sequence alignment is aligned with all training sequences based on local sequence alignment. The alignment process is repeated several times with different score matrices as shown in Table 2. The outputs of all

Table 1 Distribution of the sequences across different classes in the SWISS-PROT enzyme database

Class	Number of sequences	Percentage (%)
Oxidoreductase	1079	15.6
Transferase	2338	33.8
Hydrolase	2536	36.6
Lyase	258	3.7
Isomerase	171	2.5
Ligase	541	7.8
Total	6923	100

alignment processes are represented by ranked lists. All ranked lists are then combined to obtain a final ranked list. The final decision or the enzyme function is the element's class of the highest rank in the final ranked list. In this experiment, the distribution of the sequences in all classes are completely different (the samples of the first three classes represent 86 % of all dataset samples), which will hence favour the classes that have larger number of samples at the cost of lower performance for the other classes. The obtained results are summarised in Tables 2 and 4.

To avoid the effect of the unbalanced distribution of the class samples, each class should have approximately the same number of samples. Therefore, in our second scenario, all classes have the same number of samples, 171 samples (the number of samples of the smallest class). The results obtained are summarised in Tables 3 and 4.

Table 2 Accuracy of all single pairwise alignments based on different score matrices and the proposed fusion model using the unbalanced dataset

	Number of sequences	SA1	SA2	SA3	SA4	SA5	SA6	SA7	Proposed model
C1	540	490	489	482	486	490	493	492	508
C2	1170	1054	1055	1046	1051	1060	1060	1064	1107
C3	1270	1170	1172	1162	1162	1163	1166	1166	1213
C4	130	101	101	98	100	109	111	112	106
C5	90	65	66	62	63	70	69	68	69
C6	270	236	234	235	237	241	244	248	242
Total	3470	3116 89.8 %	3119 89.9 %	3085 88.9 %	3099 89.3 %	3133 90.3 %	3147 90.7 %	3154 90.9 %	3245 93.5 %

Table 3 Accuracy of all single pairwise alignments based on different score matrices and the proposed fusion model using the balanced dataset

	Number of sequences	SA1	SA2	SA3	SA4	SA5	SA6	SA7	Proposed model
C1	171	144	148	147	145	145	146	145	150
C2	171	162	161	160	158	156	159	162	162
C3	171	162	166	163	160	160	162	165	162
C4	171	152	150	149	149	145	151	151	150
C5	171	141	146	144	137	137	139	140	142
C6	171	150	153	149	140	149	149	153	158
Total	1026	911 88.8 %	910 88.7 %	912 88.9 %	889 86.7 %	892 86.9 %	906 88.3 %	916 89.3 %	928 90.5 %

Table 4 A comparison between single sequence alignment results and the fusion of all ranked lists that resulted from different sequence alignment processes based on different score matrices

Type score matrix	First experiment (unbalanced data)	Second experiment (balanced data)
BLOSUM62	89.8	88.8
BLOSUM100	89.9	88.7
BLOSUM30	88.9	88.9
PAM10	89.3	86.7
PAM100	90.3	86.9
DAYHOFF	90.7	88.3
GONNET	90.9	89.3
Fusion of all sequences (proposed model)	93.5	90.5

4.3 Discussion

As shown in Table 2, our proposed fusion model achieves results that better than all results based on single pairwise sequence alignments. This improvement results from combining the different outputs of single pairwise alignments. The accuracy for the first three classes ranges from 89.3 to 95.5 % in all single pairwise alignment and fusion methods. On the other hand, the accuracies for the other three classes is much lower, ranging from 68.9 to 86.2 %, which reflects the effect of the unbalanced sample distribution. This was indeed the reason for the second, balanced experiment.

As we can see in Table 3, the results of single sequence alignments here ranges from 86.7 to 89.3 %, while our proposed model achieves 90.5 %. Furthermore, the accuracies for all classes now range from 85.4 to 94.7 %, which reflects that the misclassification errors are more distributed among all classes. Table 4 shows that the overall accuracy of the unbalanced data experiment are better than that of the balanced data, which is due to the class imbalance and the smaller data set.

In summary, our proposed fusion model of all single alignments is shown to achieve results that are significantly better than all single pairwise alignments, and thus provides a useful approach to enzyme function classification.

5 Conclusions

In this paper, an enzyme sequence function classification model based on ranking aggregation is proposed. In our model, the outputs of seven pairwise local sequence alignments, based on different score matrices and represented by ranked lists, are combined, using the Borda count aggregation method, to obtain a final ranked list. The function or class label of the unknown sequence is then the function of the highest ranked element. Our proposed model is shown to achieve a classification accuracy that outperforms the results obtained from individual local sequence alignment methods and also outperforms our results in [10]. Futhermore, we investigated the issue of class imbalance and showed that more consistent results are achieved by appropriate sampling.

References

1. Bairoch, A.: The enzyme database in 2000. J. Nucl. Acids Res. 28(1), 304-305 (2000). doi:10.1093/nar/28.1.304, http://www.expasy.ch/enzyme/
2. Faria, D., Ferreira, A.E., Falcão, A.O.: Enzyme classification with peptide programs: A comparative study. J. BMC bioinform. **10**(1), 1–9 (2009)
3. Gader, P.D., Mohamed, M.A., Keller, J.M.: Fusion of handwritten word classifiers. Pattern Recognit. Lett. **17**(6), 577–584 (1996)
4. Huang, C.H., Chou, S.Y., Ng, K.L.: Improving protein complex classification accuracy using amino acid composition profile. J. Comput. biol. med. **43**(9), 1196–1204 (2013)
5. Ibrahim, A., Tharwat, A.: Biometric authenticationmethods based on ear and finger knuckle images. Int. J. Comput. Sci. Issues (IJCSI) **11**(3) (2014)
6. Lee, B.J., Shin, M.S., Oh, Y.J., Oh, H.S., Ryu, K.H.: Identification of protein functions using a machine-learning approach based on sequence-derived properties. Prot. Sci. **7**(1), 27 (2009)
7. Lu, L., Qian, Z., Cai, Y.D., Li, Y.: Ecs: An automatic enzyme classifier based on functional domain composition. J. Compu. Biol. Chem. **31**(3), 226–232 (2007)
8. Mohammed, A., Guda, C.: Computational approaches for automated classification of enzyme sequences. J. Prot. Bioinform. **4**, 147–152 (2011)
9. Sarac, O.S., Gürsoy-Yüzügüllü, Ö., Cetin-Atalay, R., Atalay, V.: Subsequence-based feature map for protein function classification. J. Comput. Biol. Chem. **32**(2), 122–130 (2008)
10. Sharif, M.M., Tharwat, A., Amin, I.I., Ella Hassanien, A., Hefeny, H.A.: Enzyme function classification based on sequence alignment. In: Information Systems Design and Intelligent Applications, pp. 409–418. Springer (2015)
11. Syed, U., Yona, G.: Enzyme function prediction with interpretable models. In: Computational Systems Biology, pp. 373–420. Springer (2009)
12. Tharwat, A., Gaber, T., Hassanien, A.E.: Cattle identification based on muzzle images using gabor features and svm classifier. In: Advanced Machine Learning Technologies and Applications, pp. 236–247. Springer (2014)

13. Tharwat, A., Ibrahim, A.F., Ali, H.A.: Multimodal biometric authentication algorithm using ear and finger knuckle images. In: Proceedings of Seventh International Conference on Computer Engineering & Systems (ICCES). pp. 176–179. IEEE (2012)
14. Tian, W., Skolnick, J.: How well is enzyme function conserved as a function of pairwise sequence identity? J. Mol. Biol. **333**(4), 863–882 (2003)
15. Xiong, J.: Essential Bioinformatics, p. 1. Cambridge University Press, Cambridge (2006)
16. Young, H.P.: Condorcet's theory of voting. Am. Polit. Sci. Rev. **82**(04), 1231–1244 (1988)

Mining of Frequent Action Rules

Agnieszka Dardzinska and Anna Romaniuk

Abstract An action rule is constructed as a series of changes, or actions, which can be made to some of the flexible characteristics of the information system that ultimately triggers a change in the targeted attribute. The existing action rules discovery methods consider the input decision system as their search domain and are limited to expensive and ambiguous strategies. In this paper, we define and propose the notion of action base as the search domain for actions, and then propose a strategy based on the FP-Growth algorithm to achieve high performance in action rules extraction. This method was initially tested on real medical diabetic database. The obtained results are quite promising.

Keywords Action rules · Association · Action base · FP-growth · Frequent action tree · Decision system · Information system · Diabetic

1 Introduction

Action rules were first introduced in [10] as a new class of rules that provides hints on possible actions a user should take to achieve a desired goal. Mining action rules is defined as the process of identifying patterns in a decision system capturing the possible changes to certain object attributes that may lead to a change in the decision value [2, 10]. Generally, action rule mining operates on a decision system [2, 7, 8] with objects having three classes of attributes: stable or semi-stable, flexible and decisions. The stable attributes are attributes that cannot be changed or, in some approaches, require a very high cost to change them [2, 5, 7]. Taking into consideration medical database with diabetic diseases, examples of stable attributes are date of birth, sex, age, SSN. Simultaneously, flexible attributes are attributes which

A. Dardzinska (✉) · A. Romaniuk
Department of Mechanics and Computer Science, Bialystok University of Technology,
Ul. Wiejska 45a, 15-351 Bialystok, Poland
e-mail: a.dardzinska@pb.edu.pl

A. Romaniuk
e-mail: a.romaniuk@doktoranci.pb.edu.pl

D. Ryżko et al. (eds.), *Machine Intelligence and Big Data in Industry*,
Studies in Big Data 19, DOI 10.1007/978-3-319-30315-4_8

values can change, such as blood glucose, blood pressure, health condition, etc. The decisions are attributes which the user would like to change, from one state to more desirable. An example would be the medical treatment kind of illness.

Existent action rules discovery methods use a decision table as their primary search domain. In our approach, the discovery of action rules is based on a domain of actions that we create from the decision system, called the action base. The main contribution of this paper is to present action rules mining problem as the association mining problem framework using the action base as the new search domain.

2 Preliminaries

We assume that $S = (X, A, V)$ is an information system, where:

- X is a nonempty, finite set of objects,
- A is a nonempty, finite set of attributes,
- $V = \{\cup V_a : a \in A\}$ is a set of all attributes values.

Additionally, $a : X \to V_a$ is a function for any $a \in A$, that returns the value of the attribute of a given object. The attributes are divided into different categories: set of stable attributes A_{St}, set of flexible attributes A_{Fl} and set od decision attributes D, such that $A = A_{St} \cup A_{Fl} \cup D$. In this paper we analyze information systems with only one decision attribute d. The example of an information system S is represented as Table 1.

Information system is represented by eight objects, one stable attribute a, two flexible attributes b, c and one decision attribute d.

Action rules, with definition presented in Sect. 3 of this paper, are very interesting and promising in medical treatment fields. They can be extracted from a decision system that describes a possible transition of objects from one state to another with respect to a distinguished attribute called a decision attribute. Previous methods based on extracting action rules [2, 7, 8, 10–12] were based on an existing set of classification rules. Certain pairs of these rules were combined to assign objects

Table 1 Information system S

X	a	b	c	d
x_1	a_1	b_1	c_2	d_1
x_2	a_1	b_1	c_1	d_1
x_3	a_2	b_1	c_1	d_1
x_4	a_2	b_2	c_1	d_2
x_5	a_2	b_2	c_2	d_2
x_6	a_2	b_1	c_1	d_1
x_7	a_2	b_2	c_1	d_2
x_8	a_2	b_1	c_2	d_2

from one class to another. There is also a method which allows to explore action rules directly from the decision system [7]. In [8, 9, 12], the proposed algorithm, called Action Rules Discovery (ARD), builds rules for a given decision using an iterative marking strategy. It considers the change in attribute value as an atomic-action-term of length one, and then an action-term is a composition of atomic-action-terms. ARD starts by generating all atomic-action-terms for a given set of attribute values and assigning a mark (unmarked, positive, negative) based on standard support and confidence measures. The action-terms marked as positive are used to construct the action rules. The unmarked terms are placed into the list. Next, it generates all possible action-terms of length two by combining terms in the list. The process continues iteratively, creating terms of greater length, until the fixed point is reached. In [8] authors presented an association type of action rules and used an Apriori like strategy to find frequent action sets to induce action rules. Like ARD, the algorithm AAR (Association Action Rule) considers atomic action sets being the fine granule used to construct longer rules (similar to items and item sets in association mining). The Apriori algorithm is used with few modifications. The main changes are mostly based on modifications to the definition of support and confidence and the calculation of the measures directly from the input decision system. Although these approaches have different definitions for objective measures like support and confidence, they use the same idea of atomic-action set, action set and Standard Interpretation.

3 Action Rules

By atomic action set we define the expression $(a, a_1 \rightarrow a_2)$ where a is an attribute in A and a_1, a_2 are values of a. If the attribute is stable or did not change its value then the atomic action set is expressed as (a, a_1). The domain of an atomic action set is its attribute.

Example 1 Consider *Diet* is a flexible attribute with values $V_D = \{1000\,\text{kcal}, 1200\,\text{kcal}, 3000\,\text{kcal}\}$. The atomic action set $(Diet, 3000\,\text{kcal} \rightarrow 1200\,\text{kcal})$ means changing the value of attribute *Diet* from 3000 to 1200 kcal.
Action sets are constructed as the conjunction of atomic actions. If t_1, t_2 are two atomic action sets with different attributes, then $t = t_1 * t_2$ is an action set. The domain of action set t is the set of attributes from all its atomic action sets $Dom(t) = Dom(t_1) \cup Dom(t_2)$.

Example 2 Let us assume that *Age* is a stable attribute with values. $V_A = \{20, 40, 60\}$ and *Weight* is a flexible attribute with values $V_W = \{overweight, equal\_weight, underweight$. An action set could be the composition $(Age, 40) * (Weight, overweight \rightarrow equalweight) * (Diet, 3000\,\text{kcal} \rightarrow 1200\,\text{kcal})$ means: for patients of Age 40, lower the body weight and change the diet. The introduction of the Standard Interpretation is the basis of measures like support and confidence [1, 2, 7]. In association mining, the support of an itemset is the count of all objects in information system. For action rules, we need to take into consideration two sets. The first set is the set of all the objects with attributes value equal to the initial state of the action. The second one is the set of all the objects having attributes values equal to the values of the final state of the action.

Example 3 The Standard interpretation of action set $Ns[(Age, 40) * (Weight, overweight \rightarrow equalweight) * (Diet, 3000\,\text{kcal} \rightarrow 1200\,\text{kcal})] = [\alpha 1, \alpha 2]$, where:
$\alpha 1 = \{x \in X : [Age(x) = 40] * [Weight(x) = overweight] * [Diet, 3000\,\text{kcal}]\}$
$\alpha 2 = \{x \in X : [Age(x) = 40]*[Weight(x) = equalweight]*[Diet, 1200\,\text{kcal}]\}$.
Action rule r is expressed as an implication $r = [t1 \rightarrow t2]$, where $t1$ and $t2$ are two action sets.
Assume that t is an action set with standard interpretation $Ns(t) = [\alpha 1, \alpha 2]$. The support of this action set t is defined as in ARD strategy:

$$Sup(t) = min\{card(\alpha 1), card(\alpha 2)\}.$$

The confidence of an action *rule* $r = [t1 \rightarrow t2]$, where $Ns(t1) = [\alpha 1, \alpha 2]$ and $Ns(t2) = [\omega 1, \omega 2]$, with $\alpha 1 \neq 0$ and $\alpha 2 \neq 0$ is defined as follows:

$$Conf(r) = \frac{card(\alpha_1 \cap \omega_1)}{card(\alpha_1)} . \frac{card(\alpha_2 \cap \omega_2)}{card(\alpha_2)}$$

Generating action rules is similar to association rule mining where frequent item sets are first extracted. The algorithm, which is based on Apriori, generates actions sets with support that exceeds specified two threshold values: minimum support λ_1, and minimum confidence λ_2. Any action set that meets this criterion is a frequent action set. An action rule is constructed and tested as following:

If $t$is a frequent action set and *t1 is* a subset of t then $r = [t - t1 \rightarrow t1]$.

$$If\, Sup(r) < \lambda_1 \text{ and } Conf(r) < \lambda_2, \text{ then } r \text{ is ruled out.}$$
$$If\, Sup(r) \geq \lambda_1 \text{ and } Conf(r) < \lambda_2, \text{ then } r \text{ is not marked.}$$
$$If\, Sup(r) \geq \lambda_1 \text{ and } Conf(r) \geq \lambda_2, \text{ then } r \text{ is a valid rule.}$$

Presented method can generate a large number of rules. However the process does not constrain what the decision attribute can be, it does not require a decision attribute to be specified at all. Moreover, if the user is interested in a particular change object from one class to another like $(Weight, overweight \rightarrow equalweight)$, there is no guarantee that we will generate the required rules. For instance, if minimum support is 10, and (*Weight, overweight*) has support equal 9, then no rules containing (*Weight, overweight*) would be generated.

Methods which generate action rules mainly work in information systems in form of typical decision table, where there is a set of objects, set of attributes, and set of values of attributes [6]. As a result, forming flexible *atomic action sets* is revealed and connected with the action set generation. Hence, one would need to calculate the support of (*Weight, overweight*) and (*Weight, equal weight*) first, in order to obtain the support of ($Weight, overweight \rightarrow equalweight$). This operation needs to be performed for all iterations. The first iteration generates frequent action sets composed of 1 element, the second iteration generates frequent action sets composed of 2 elements, the third iteration generates frequent action sets composed of 3 elements, and so on, until the fixed point is reached. In such way we obtain knowledge base. The information stored in form of action table are useful in later reclassification process, where objects move from one class to another.

Let us assume that Table 1 is the decision table and the decision attribute is $(d, d_1 \rightarrow d_2)$. It can be divided into two sub tables *Class1* and *Class2*, where:
Class1 = $\{x \in X : d(x) = d_1\}$, i.e. all the objects with decision value d_1.
Class2 = $\{x \in X : d(x) = d_2\}$, i.e. all the objects with decision value d_2.
$\forall x \in Class1, \forall y \in Class2 : (x \rightarrow y)$ is a possible transition describing a new action set *t*,
$\forall a \in A$: if a is flexible and $a(x) \neq a(y)$ then $a(t) = a(x) \rightarrow a(y)$,
$\forall a \in A$: if a is stable and $a(x) \neq a(y)$ then $a(t)$ is discarded,
$\forall a \in A$: if $a(x) = a(y)$ then $a(t) = a(x)$.
The action base, which will be formed, will contain the necessary action sets which help to move every object from *Class1* to *Class2*. The action base will contain *card*(*Class1*) * *card*(*Class2*) action sets. Each row in the action base reflects the necessary action sets for moving an object from d_1 to d_2.

Example 4 To generate action base we start from the decision Table 1 and organize it into two sub tables, as shown in Table 2, with respect to the decision attribute d. The action base includes change values as part of attribute value. The decision cardinality $card(\{d, d_1 \rightarrow d_2\})$ is equal to the number of rows in the action base. Moreover, the action base does not allow us to recover the individual counts of $\omega 1$ or $\omega 2$ nor the individual counts of $\alpha 1$ or $\alpha 2$. This information is not needed, then support for t, and confidence for r, where $t = t1 \cdot t2$ and $t2 = \{d, d_1 \rightarrow d_2\}$, are given by:
$Sup(t) = card(\alpha 1 \cap \omega 1) \cdot card(\alpha 2 \cap \omega 2), Conf(r) = \frac{Sup(r)}{Sup(t1)}$
Above description are the same support and confidence measures used in the traditional Association Mining [1, 4].

Table 2 Information system*S* divided into two separate classes

X	a	c	c	d	
x_1	a_1	b_2	c_2	d_1	*Class1*
x_2	a_1	b_1	c_1	d_1	
x_3	a_2	b_1	c_1	d_1	
x_6	a_2	b_1	c_1	d_1	
x_4	a_2	b_2	c_1	d_2	*Class2*
x_5	a_2	b_2	c_2	d_2	
x_7	a_2	b_2	c_1	d_2	
x_8	a_1	b_1	c_2	d_2	

The problem is that the *Sup(r)* can be directly calculated from the action base, but not the $Sup(t1)$. This is because the action base does not contain all occurrences of $\alpha 1$ and $\alpha 2$. However, the support for $\alpha 1$ and $\alpha 2$ can be easily calculated from the decision table. Formally, the support of an action set $t1$ with standard interpretation $Ns(t1) = [\alpha 1, \alpha 2]$ can be determined as $Sup(t1) = card(\alpha 1) \cdot card(\alpha 2)$.

4 Frequent Action Rules

In this paper we will use FP-Growth algorithm [2] to form frequent action tree, to generate the frequent action sets and then frequent action rules. Frequent action tree is obtained by order the transaction attributes values by their frequency, prune those that do not exceed minimum support given by user, and then insert action sets into a form of a tree. The result is compact data structure. The process of extracting action rules using frequent action tree is as follows:

1. Generate all the atomic action sets from set of attributes.
2. Calculate the frequency and the support of each atomic set.
3. Build and reorder the action base.
4. Build the frequent action tree from the action base.
5. Extract frequent action sets.
6. Extract action rules.

To better explain our strategy, we go through the example on the decision table described in Table 1. We use $\lambda_1 = 2$ and $\lambda_2 = 0.75$ as a minimum support and minimum confidence, respectively. Our goal is to reclassify objects from class d_1 to d_2. First we have to generate the atomic sets. These are all the possible transactions for each attribute.

$$AS = \{(a, a_1), (a, a_2), (b, b_1), (b, b_2), (b, b_1 \rightarrow b_2), (b, b_1), (c, c_1), (c, c_2),\\ (c, c_2 \rightarrow c_1)\}$$

For each atomic set we calculate its frequency. We scan the whole decision table and count the occurrence of each attribute with respect to decision (d_1, d_2).

All remain atomic action set build action base, which is in form of previous Table 3. We put all atomic set into descending order:

$$\{(a, a_2), (b, b_1 \rightarrow b_2), (c, c_1 \rightarrow c_2), (b, b_1), (c, c_1), (b, b_2), (c, c_2)\}$$

These ordered action sets form new action base.
To build the frequent action tree we insert each action set, from the action base, into the tree (Fig. 1) using the atomic action sets as the nodes. Each time a node is inserted or reused we increment its weight (Table 4).
Frequent action sets obtained from Table 5 are: $(a_2, b_1 \rightarrow b_2, c_1 \rightarrow c_2)$, $(a_2, b_1 \rightarrow b_2, c_1)$, $(a_2, b_1, c_1 \rightarrow c_2)$, $(a_2, b_1 \rightarrow b_2)$.

Table 3 Action base for decision attribute d

	a	b	c
$x_1 \to x_4$	–	b_2	$c_2 \to c_1$
$x_1 \to x_5$	–	b_2	c_2
$x_1 \to x_8$	–	$b_2 \to b_1$	c_2
$x_2 \to x_4$	–	$b_1 \to b_2$	c_1
$x_2 \to x_5$	–	$b_1 \to b_2$	$c_1 \to c_2$
$x_2 \to x_8$	–	b_1	$c_1 \to c_2$
$x_3 \to x_4$	a_2	$b_1 \to b_2$	c_1
$x_3 \to x_5$	a_2	$b_1 \to b_2$	$c_1 \to c_2$
$x_3 \to x_8$	a_2	b_1	$c_1 \to c_2$
$x_6 \to x_4$	a_2	$b_1 \to b_2$	c_1
$x_6 \to x_5$	a_2	$b_1 \to b_2$	$c_1 \to c_2$
$x_6 \to x_8$	a_2	b_1	$c_1 \to c_2$

Fig. 1 Frequent action tree

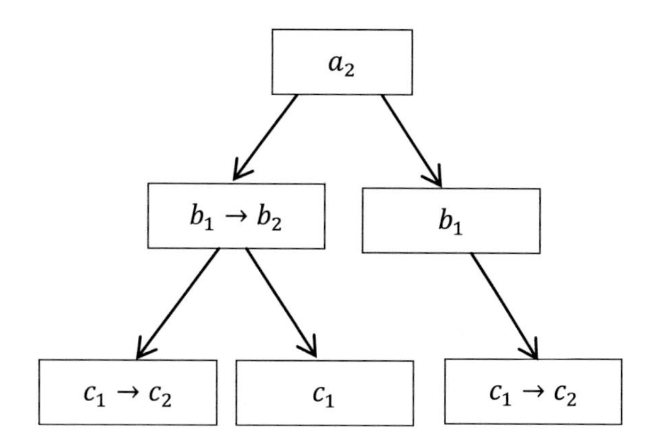

Table 4 Support of atomic action set

Atomic action set	Support of atomic action set
(a, a_1)	0 not marked
(a, a_2)	6
(b, b_1)	3
(b, b_2)	2
$(b, b_1 \to b_2)$	6
$(b, b_2 \to b_1)$	1 not marked
(c, c_1)	3
(c, c_2)	2
$(c, c_1 \to c_1)$	6
$(c, c_2 \to c_1)$	1 not marked

Table 5 New ordered action base for decision attribute d

	a	b	c
$x_3 \rightarrow x_4$	a_2	$b_1 \rightarrow b_2$	c_1
$x_3 \rightarrow x_5$	a_2	$b_1 \rightarrow b_2$	$c_1 \rightarrow c_2$
$x_3 \rightarrow x_8$	a_2	b_1	$c_1 \rightarrow c_2$
$x_6 \rightarrow x_4$	a_2	$b_1 \rightarrow b_2$	c_1
$x_6 \rightarrow x_5$	a_2	$b_1 \rightarrow b_2$	$c_1 \rightarrow c_2$
$x_6 \rightarrow x_8$	a_2	b_1	$c_1 \rightarrow c_2$

From frequent action sets we get frequent action rules:
$r1 = (a, a2) \cdot (b, b_1 \rightarrow b_2) \cdot (c, c_1 \rightarrow c_2) \rightarrow (d, d_1 \rightarrow d_2)$ with $conf(r1) = 0.33$
$r2 = (a, a2) \cdot (b, b_1 \rightarrow b_2) \cdot (c, c_1) \rightarrow (d, d_1 \rightarrow d_2)$ with $conf(r2) = 0.66$
$r3 = (a, a2) \cdot (b, b_1) \cdot (c, c_1 \rightarrow c_2) \rightarrow (d, d_1 \rightarrow d_2)$ with $conf(r4) = 1$
$r4 = (a, a2) \cdot (b, b_1 \rightarrow b_2) \rightarrow (d, d_1 \rightarrow d_2)$ with $conf(r5) = 0.66$
For each extracted rule we calculate its confidence. In our example only $r3$ exceeds the minimum confidence for a valid association action rule extracted from information system S.

5 Experiments

Our dataset contains clinical data of 150 patients affected by diabetes. Patients are characterized by 15 attributes (5 stable and 10 flexible), and classified into three groups: diabetes, pre-diabetes and healthy. The goal was to find rules which help to reclassify patients to group of healthy persons.:
$[class, diabetes \rightarrow healthy], [class, pre\text{-}diabetes \rightarrow healthy]$
We assumed $\lambda_1 = 30\,\%$, $\lambda_2 = 70\,\%$, and we obtained several action rules. Some of them are given below:

- If patient with type 2 diabetes is overweight and lose weight and begin regular physical activity, then his blood glucose returns to normal (sup = 70 %. Conf = 84 %)
- If patient will parasite cleanser, then his blood glucose will return to normal (sup = 35 %. Conf = 72 %)
- If patient will stop all hormone-effecting medications during the diabetes curing process, then her glucose tolerance will increase (sup = 90 %. Conf = 95 %)
- If patient will eliminate all processed foods, white foods, artificial sweeteners, tap water, artificial additives, then his blood glucose will remain normal (sup = 80 %. Conf = 85 %).

6 Conclusions and Future Work

In this paper, we propose the action base for frequent action rules mining. The action base transforms the complex problem of finding action rules from a decision table, into finding action rules from an action base. Applying this method to real medical databases we can notice that it is quite useful method, giving promising results. In the future direction, we will try to modify the method to compact frequent action set. We will also work on information systems with several decision attributes.

References

1. Agrawal, R., Srikant, R.: Fast algorithm for mining assocation rules. Int. Conf. Very Larg. Data Bases pp. 487–499 (1993)
2. Dardzinska, A.: Action Rules Mining. Springer, Berlin (2013)
3. Deogun, J., Raghavan, V., Sever, H.: Rough set based classification methods and extended decision tables. In: International Workshop on Rough Sets and Soft Computing. pp. 302–309 (1994)
4. Han, J., Pei, J., Yin, Y.: Mining frequent patterns without candidate generation, ACM SIGMOD. In: International Conference on Management of Data. pp. 1–12 (2000)
5. Im, S., Ras, Z.: Action Rule Extraction from a Decision Table: ARED. In: International Syposium on Methodologies for Intelligent Systems pp. 160–168 (2008)
6. Pawlak, Z.: Information systems—theoretical foundations. Inf. Syst. J. Elsevier **6**, 205–218 (1981)
7. Ras, Z., Dardzinska, A.: Action Rules Discovery without Pre-existing Classification Rules. In: The Sixth International Conference on Rough Sets and Current Trends in Computing pp. 181–190 (2008)
8. Ras, Z., Dardzinska, A., Tsay, L., Wasyluk, H.: Association Action Rules. In: IEEE International Conference on Data Mining Workshops pp. 283–290 (2008)
9. Ras, Z., Tsay, L.: Discovering Extended Action Rules (System DEAR). In: International IIS IIPWM'03 Conference. pp. 293–300 (2003)
10. Ras, Z., Wieczorkowska, A.: Action-Rules: How to Increase Profit of a Company, The Fourth European Conference on Principles and Practice of Knowledge Discovery in Databases. pp. 587–592
11. Ras, Z., Wyrzykowska, E., Wasyluk, H.: ARAS: Action Rules Discovery Based on Agglomerative Strategy. In: Third International Workshop on Mining Complex Data. pp. 196–208 (2007)
12. Tsay, L., Ras, Z.: Action rules discovery: system DEAR2, method and experiment. J. Exp. Theor. Artif. Intell. pp. 119–128 (2005)

6 Conclusions and Future Work

[illegible]

References

[illegible]

Part III
Text and Multimedia Processing

Automatic Translation of Multi-word Labels

Grzegorz Protaziuk, Marcin Kaczyński and Robert Bembenik

Abstract Application of semantic resources often requires linking phrases expressed in a natural language to formally defined notions. In case of ontologies lexical layers may be used for that purpose. In the paper we propose an automatic machine translation method for translating multi-word labels from lexical layers of domain ontologies. In the method we take advantage of Wikipedia and dictionaries services available on the Internet in order to provide translations of thematic texts from a given area of interest. Experimental evaluation shows usefulness of the proposed method in translating specialized thematic dictionaries.

Keywords Domain label translation · Automatic translation · Wikipedia application

1 Introduction

Semantic resources such as ontologies are becoming more and more important in various information systems nowadays. Domain ontologies in IT are considered to be formal descriptions of selected pieces of the world enabling various beings to share common understating of those areas of interest. The common understanding concerns a conceptual layer of ontologies defining semantics. However, in many situations where text written in a natural language is involved, associations between phrases expressing a certain notion in the natural language and a formal definition of that concept stored in an ontology are needed. Lexical layers of ontologies may be perceived as a kind of bridge [11] linking phrases in a natural language to an ontology concept, but usually a newly built ontology has a lexical layer expressed

G. Protaziuk (✉) · M. Kaczyński · R. Bembenik
Institute of Computer Science, Warsaw University of Technology,
Nowowiejska 15/19, 00-665 Warszawa, Poland
e-mail: G.Protaziuk@ii.pw.edu.pl

M. Kaczyński
e-mail: M.Kaczynski@elka.pw.edu.pl

R. Bembenik
e-mail: R.Bembenik@ii.pw.edu.pl

D. Ryżko et al. (eds.), *Machine Intelligence and Big Data in Industry*,
Studies in Big Data 19, DOI 10.1007/978-3-319-30315-4_9

only in one or few languages. The practical usefulness of such an ontology is more or less limited. Manual translation is a costly and time consuming process. Also, such translation may require involvement of a domain specialist as it is quite probable that a domain ontology includes definitions of very specialized concepts.

This paper addresses that problem and proposes an approach to automatic generation of possible translations of labels included in a lexical layer of an ontology. The introduced method uses external resources available on the Internet and returns phrases taken from thematic documents as the proposed translations.

The problem of translating lexical layers of ontologies is related to the general problem of automatic machine translation, particularly translation of dictionaries. Source data for algorithms of automatic translation can be perceived twofold, as (1) a multilingual comparable corpus, a set of collections of texts referring to similar topics, and (2) a parallel corpus, a set of collections of linked texts being adjusted translations. In methods dealing with parallel bilingual corpora [1, 6, 8, 13] usually firstly pairs of corresponding sentences from texts are identified, and then pairs of corresponding words/phrases are determined for the sentences paired in the previous step. Typically, such an adjustment is performed by means of a probabilistic model of languages proposed by authors of a translation method. Examples of application of statistical machine translations methods can be found e.g. in [4]. Also, a lot of methods have been introduced in the literature, in which a statistical model is enhanced by various linguistic information or semantics. For example, in [3] the source-context similarity approach to creation of a statistical model was applied. In that method a translation unit consists of three elements: phrase-source-side, phrase-target-side, and source-context (a source sentence from which a translation unit was originally extracted). The source-context is used for calculation of similarity between an input sentence to be translated and a translation unit. In [2] the synchronous context-free grammars (SCFGs) for creating statistical models of translation were used. Authors incorporated modality and named entity tags as higher-order symbols in translations rules. A review of such methods can be found in [5, 14]. Methods of statistical machine translation can provide very high precision of translations but (1) require parallel corpora that are rather rare and (2) taking into consideration changes in languages requires rebuilding of language models.

In methods dealing with comparable corpora it is usually assumed that words often occurring together in texts in a source language also often occur together in texts in a target language. Such relation between two words A and B takes place if a distance (number of words) between A and B in a given text is not greater than a certain threshold. In these methods also a base dictionary including translations of selected words is applied for comparing contexts of words and for building a resultant dictionary. In [7, 12] such dictionary is an input parameter, whereas in [9, 10] it is created from scratch. A dictionary is built based foremost on identical spelling of some words in both languages and usage of rules which eliminate typical differences in spelling of words in the considered languages. The reported results obtained by means of such methods were essentially worse than from methods using parallel corpora. However, they may be applied in many more situations because comparable corpora are significantly more common (e.g. Wikipedia) than the parallel ones.

The rest of the paper is structured as follows. Section 2 presents our approach to labels translation, Sect. 3 contains a formal description of the proposed method, Sect. 4 discusses results of the experiments, and, finally, Sect. 5 concludes the paper.

2 The Translation Method

The proposed method of translation is aimed at translating multiword labels from a lexical layer of a domain ontology. In our approach the focus is set on finding correct translations coming from vocabulary used in a given thematic area. We search for translations in a text document repository and assume that proper translations of labels occur in those texts. In our method external services are used to obtain data applied for building proper translations. We use data from Wikipedia both for verification of correctness of the prepared translations as well as for selecting the most suitable translations for a given domain of interest. We use Wikipedia also for finding translations.

We begin the presentation of a proposed approach from a short description of Wikipedia from the perspective of our method and then we present in detail the proposed translation procedure.

2.1 Wikipedia

Wikipedia is an online, publicly accessible, multilingual encyclopedia which accumulates knowledge of many different fields. It has a multi-level structure of references and categories. The categories may be perceived as the domain identifiers and a hierarchy of categories may be used for selecting pages from a given domain (more or less general). We use categories for building a domain-specific repository of articles. In Wikipedia there are three types of links, namely: internal, external, and inter-lingual. The last ones indicate links connecting articles written in different languages describing the same notions. This makes it possible to use these links for finding correct translations. Moreover, it is easy to determine categories (domains) associated with inter-lingual links which, in turn, allows checking if a found translation refers to a given thematic area. In the method we also take advantage of disambiguation pages grouping synonyms and the redirection mechanism used in Wikipedia to represent the issues that occur with several different names.

2.2 Translation Procedure

The proposed method of label translations consists of the three following ways of translation:

1. *Translation based on Wikipedia* in which we take advantage of inter-lingual links and the redirection mechanism.
2. Translation utilizing *the TransLab* procedure, which consists of the following steps:

   - translations of single words—obtaining possible translations for each word included in a label;
   - generation of candidate translations of multi-word labels;
   - verification of the candidate translations.

3. Translation based on translator services

   - generating candidate translations based on Internet translator services;
   - verification of candidate translations.

Repository. A text document repository used in the translation method may be any set of thematic documents written in a target language. It is assumed that translated labels and documents in a repository concern the same domain area. In our implementation texts are retrieved from Wikipedia based on keywords (which are equated with names of Wikipedia categories) provided by a user. We use the list of pages associated with Wikipedia categories to obtain the needed articles. Also, we recursively retrieve articles linked with subcategories of the previously searched categories.

Translation based on Wikipedia. In this method we search for pages whose titles are the same as the translated label. If such a page (or page found by means of the Wikipedia redirection mechanism) has an inter-lingual link to a page written in a target language the title of a linked page is treated as a translation. As in our research we focused on translating labels from English into Polish and the most English titles of pages in Wikipedia are composed of words in their singular forms, we additionally applied the following transformations of words included in the labels:

- removing suffix 's', e.g. computers → computer;
- changing suffix 'ies' into 'y', e.g. cities → city;
- changing suffix 'ves' into 'f' and 'fe', e.g. (halves → half, knives → knife),
- changing suffix 'es' into 'e' and removing that suffix, e.g.: bridges → bridge, dresses → dress.

As the applied method of transformation is very simple and it does not take into account the numerous exceptions that occur in English it does not guarantee generation of the correct singular forms of words. Generally, usage of that method should give good results, but we cannot exclude situations in which usage of the method may result in incorrect translations. As we build a domain repository from Wikipedia pages we can perform a simple verification of the found translations, namely: a given translation is valid only if it is taken from a page included in the domain repository.

Translation of single words. The possible translations of each word included in a label are retrieved by means of machine translation services available on the Internet. For this purpose a translated label is split into single words, and with each such word a list of the found translations is associated. In our experiments we used the following services: *Bing Translator* (www.bing.com/translator), *Google Translate* (http://google.translate.pl), *1 A Tradovium* (http://www.1a.biz.pl/slownik), and Translate.pl (http://translate.pl).

Generation of candidate translations. For generating candidate translations for labels single words are used, namely: for a label consisting of n words all possible sequences of terms $t_1, t_2, \ldots, t_n$ are created, where term t_k belongs to the list of translations associated with the word at the kth position in the translated label. As the order of words in a proper translation of a label in a target language may be different from the order of words in the translated label we generate candidate translations for all permutations of words included the translated label.

Translation based on translator services. For research purposes we implemented a method for obtaining translations of whole phrases from the Google Translate service. The main disadvantage of the service is that it returns only one version of the translation. Also, it should be noted that the service returns a translation for any label, and in the worst case a resultant translation is exactly the same as a source phrase.

Verification of the candidate translations. The aim of this step is to select translations from the set of candidate translations that are valid in a target language and are appropriate for a given thematic area. For that purpose the domain repository is used. Generally, we check whether a candidate translation occurs in text documents contained in that base; if so the translation is considered correct. Verification is performed for both methods of generating candidate translations: *TransLab* procedure and *Translation based on translator services.*

2.3 Searching Phrases in a Domain Repository

Determining occurrence of a word in texts

A proper translation of a multi-word label may include words that are not in their basic form. It especially concerns situations in which a target language is highly inflected (e.g. Latin, Polish). It is probable that by using machine translation services we do not obtain all possible grammatical forms of translations of a given word. Usually because of that such services return translations only in the basic word forms. A variety of grammatical forms of a given word is a cause of a problem with proper determination of occurrences of that word in texts stored in a domain repository. In order to address that problem we defined a *represented relation* between words, namely: a word A included in a label is in a represented relation with a word Bincluded in a text if the word B may be considered an occurrence of the word A in that text. In the sequel, we denote that relation as $\Gamma(A, B)$. We used two methods for determining $\Gamma(A, B)$:

- By using dictionary variations (inflections, conjugations, etc.) of words. Such dictionary may be seen as a set of the following pairs: a word and a list of its all possible variations. In this method two words are in a represented relation if there is a list of variations that includes those both words. (In the experiments we used a dictionary available at the address: http://sjp.pl/slownik/odmiany/.)
- By computing similarity between words. The similarity between two words (character strings) a and b is calculated by means of $sim(a, b)$ function defined in the following way:

$$sim(a,b) = \frac{1.0}{1.0 + \frac{lev(a,b)}{min(|x|,|y|)}}$$

where: lev(a, b) is the Levenshtein distance between a and b,
$|x|$—number of characters in x
The similarity of two strings is a value form the range $\langle 0, 1\rangle$. The similarity is equal to 1 if two strings are the same.

Efficiency of searching in a domain repository The number of candidate translations may be very high, e.g. if a translated label consists of three words and each word is translated into five terms and each term has 4 variants, the number of candidate translations is equal to $(5*4)^3 = 8000$. A verification of candidate translations may be performed in an efficient manner provided that a repository management system offers fast access to sentences in which a given word occurs.

3 Formal Description of the TransLab Procedure

In the sequel we use the following notation:

- W—a set of words.
- $W^{\text{lang}} / T^{\text{lang}}$—a set of words/terms from a given natural language *lang*.
- λ—label. A label is a sequence of words from a given finite set of words. $\lambda = \langle w_1, w_2, \ldots, w_n \rangle$, $w_i \in W$. A label built from k words is a *k-label.*

3.1 Translation of a Single Word

Function $wordtran_{A\rightarrow B}$ is a translation function which for a given word w from a language A returns a set of terms (possible translations of the word w) from a language B. Formally:

$$wordtran_{\mathrm{A}\rightarrow\mathrm{B}}: w^{\mathrm{A}} \rightarrow T^{\mathrm{B}},$$

where w^A—a word from a language A, $w^A \in W_{langA}$;
and T^B is a set of terms from a language B $T^B = \{t_1, t_2, \ldots, t_n\}, t_i \in W_{langB}$.

3.2 Generation of the Candidate Translations

Function $labeltran_{A \to B}$ is a translation function which for a given label consisting of words from a language A returns a set of sets of possible translations for each word included in the label. Formally:

$$labeltran_{A \to B} : \lambda^A \to \cup_{i=1}^{n} T_i^B$$

where λ^A—a label consisting of $n$words from a language A, $\lambda^A = \langle w_1, w_2, \ldots, w_n \rangle$, $w_i \in W_{langA}$

T_i^B—a set of terms from a language B, which are possible translations of a word W_i; $T_i^B = wordtran_{A \to B}(w_i)$;

$\cup_{i=1}^{n} T_i^B$—a set of sets T_i^B.

Example Given the label: *binary numeral system*

$wordtran_{E \to P}$ (binary) = {dwójkowy; podwójny; dwuskładnikowy; złożony z dwóch pierwiastków; dwuczłonowy}
$wordtran_{E \to P}$ (numeral) = {liczebnik; liczba; liczbowy; cyfrowy}
$wordtran_{E \to P}$ (system) = {system; układ; sieć; ustrój; reżim; metoda; organizm; po-rzadek; organizacja; systematyczność; formacja}
$labeltran_{E \to P}$ (binary numeral system) = {dwójkowy; podwójny; dwuskładnikowy; złożony z dwóch pierwiastków; dwuczłonowy}, {liczebnik; liczba; liczbowy;
cyfrowy},{system; układ; sieć; ustrój; reżim; metoda; organizm; po-rzadek; organizacja; systematyczność; formacja}}

For a given sequence $seqK(T^B)$ of sets T^B a candidate translation of a label $\lambda^A = \langle w_1, w_2, \ldots, w_k \rangle$ is a sequence of terms $ct = \langle t^1, t^2, \ldots, t^k \rangle$, where $t^j \in T_i^B$. As term t^i is a sequence of words a candidate translation can also be presented as a sequence of words: $ct = \langle t^1\ t^2 \ldots\ t^k \rangle = \langle w_1^1, w_2^1, \ldots, w_{j1}^1, w_1^2, w_2^2, \ldots, w_{j2}^2, \ldots, w_1^k, w_2^k, \ldots, w_{jk}^k \rangle$, where w^i constitutes a term $t^i : t^i = \langle w_1^i, w_2^i, \ldots, w_{j1}^i \rangle$. A set of candidate translations, denoted as SCT_{seqK}, is a Cartesian product of the sets of possible translations of single words: $Sct_{seqK} = T_1^B x T_2^B x \ldots x T_K^B$.

As the order of words in a proper translation of a label in a target language may be different from order of words in the translated label the final set of candidate translations, denoted as *Sct*, is a sum of candidate translations for each possible order of sets T^B; $Sct = \cup_{l=1}^{k!} Sct_{seql}$, where k is a number of words in the translated label.

3.3 Verification of Candidate Translations

In the verification phase of the method we check if a considered candidate translation occurs in some sentences taken from documents stored in a thematic repository. Formally, a candidate translation $ct = \langle tw_1\ tw_2 \ldots\ tw_k\rangle$ occurs in a sentence $s = \langle w_1 w_2 \ldots w_n\rangle$ if there exist integers $i_1,\ i_2,\ldots,\ i_k;\ i_{j+1} = i_j+1$, such that $\Gamma(tw_1,w_{i1})$, $\Gamma(tw_2,w_{i2})$, $\Gamma(tw_k,w_{ik})$.
Given:

- *RD*: a domain repository (in our case a set of articles);
- *dT*: a text document from the repository $dT \in RD$;
- *s*: a sentence (a sequence of words) $s = \langle w_1\ w_2 \ldots w_n\rangle$;

we define a candidate translation $ct = \langle tw_1 tw_2 \ldots\ tw_k\rangle$ to be a correct translation if $\exists s \subset dT,\ dT \in RD$ such that *ct* occurs in *s*. An adjusted sequence of words from the sentence *s* is added to the set of final translations. Also, for each proposed translation we provide basic statistics concerning the occurrence of that translation in the repository, namely:

- the number of occurrences in the repository (which is used to order the discovered translations);
- the number of articles containing that translation;
- the number of sentences containing that translation.

4 Experiments

In order to verify our assumptions we carried out a series of experiments. We used sets of labels which were taken from the specifications of the three ontologies presented below:

- CompSet—a set of 98 labels from the field of computer science selected in a random manner.
- ChemSet—a collection of 154 labels from the field of chemistry taken from the ontology available at http://ontology.dumontierlab.com/chemistry-primitive.
- ACMSet—a set of 1085 labels from the field of computer science and mathematics taken from the ontology ACM Computing Classification System available at http://totem.semedica.com/taxonomy/.

We evaluated the practical usefulness of the proposed procedure of translation by means of the precision measure, i.e. we calculated the percentage of correct translations in a resultant set of all translations. A translation *ct* was considered to be correct if it fulfilled one of the conditions of: (i) being equal to a reference translation *rt*, (ii) having the similarity between *rt* and *ct* greater than 0.8, (iii) for each word in *ct*

Table 1 Statistics summarizing the experiments per labels collections

	CompSet	ChemSet	ACMSet
Number of labels	98	154	1085
Labels translated correctly (first 5 proposals)	70.4 %	42.2 %	55.9 %
Labels translated correctly (first 10 proposals)	78.6 %	46.1 %	59.7 %
Labels translated correctly—all proposals	86.7 %	52.6 %	65.3 %
Labels not translated—all proposals	4.1 %	18.8 %	16.2 %
Avg. number of proposed translations	76	23	24
Avg. precision of proposed translations	0.26	0.22	0.41
Google Translator: correct translations	67.4 %	42.5	68.0
Bing Translator: correct translations	69.4 %	41.8	59.2

Table 2 Efficiency of the translation methods

Translation method	Labels set					
	CompSet		ChemSet		ACMSet	
	Found %	Correct %	Found %	Correct %	Found %	Correct %
Inter-lingual	46.9	91.3	17.5	100	29.4	99.0
Redirection	31.6	74.2	27.3	28.6	11.2	74.4
TransLab	91.8	82.2	76.6	61.7	52.7	56.3
Translators	84.7	80.7	61.0	69.2	38.2	70.5

having the similarity between words at the same position in *ct* and *rt* greater than 0.8. $sim(a, b)$ function was used to calculate the similarities.

The statistics summarizing the experiments are given in Table 1. The comparison of efficiency of different ways of translation is provided in Table 2. In this table the following notation is applied:

- *Inter-lingual*: translation based on the Wikipedia liter-ligual links;
- *Redirection*: translation based on the Wikipedia redirection mechanism;
- *Translab*: translation by means of the TransLab procedure;
- *Translators*: translation based on Internet translator services.

The achieved experimental results and their quality have been influenced not only by the quality of data returned by the used external dictionary or translator services, but also by the representativeness and quality of the documents stored in the domain repository. The complexity and the lengths of labels also had impact on the obtained results. In general, the longer and more complex labels, the worse the results.

The labels indicating concepts related to the field of computer science were composed of no more than four words and are commonly found in the texts. They are popular: about 47 % of these labels were translated using inter-lingual links. The labels of concepts related to the field of chemistry are much more complicated, which is reflected in the fact that less than 18 % of the translations were found by means of inter-lingual links. The chemical ontology also includes labels indicating very specific notions, for which it was difficult in general to find appropriate Polish translations. Many labels in the ACM ontology are composed of multiple words. One label may be a composition of two other labels referring to different concepts. They are separated by comma or joined by a conjunction. The proposed method does not ensure obtaining good results in such situations. That is because such labels seldom if ever occur in thematic texts.

In general, the proposed method of translation allowed achieving better results than the results obtained by the means of Internet translators. One exception concerns the ACM ontology for which better results were obtained using the Google translator. It can be explained by the lack of occurrences of candidate translations in thematic documents.

5 Conclusions

In this paper we presented an approach to translating multi-word labels from a lexical layer of a domain ontology. The introduced method is focused on translations from English into Polish and some detailed solutions were adjusted to realize such translations. However, the applied schema of translation is a general idea and it may be adapted for the needs of translation between other languages. In our approach we find translations used in texts concerning the areas of interest by collecting data from an external dictionary and translation services and searching for the occurrence of the generated translations in documents stored in a thematic repository. Such an approach ensures that changes in languages will be reflected in the proposed translations. The conducted experiments showed practical usefulness of our method; we were able to find many correct translations. Additionally a user will receive information about the usage frequency of a given translation for the domain articles.

References

1. Brown P., Della Pietra S., Della Pietra V., Mercer R.: The mathematics of statistical machine translation: Parameter estimation. Comput. Linguist. (1993)
2. Baker K., Bloodgood M., Callison-Burch C., Dorr B.J., Filardo N.W., Levin L., Piatko C.: Semantically-informed syntactic machine translation: a tree-grafting approach arXiv (2014)
3. Banchs R.E., Marta R. Costa-jussà M.R.: A semantic feature for statistical machine translation. In: Proceedings of the fifth workshop on syntax, semantics and structure in statistical translation. Association for Computational Linguistics (2011)
4. Bojar O., Buck C., Federmann C., Haddow B., Koehn P., Monz C., Post M., Specia L. (eds.) In: Proceedings of the Ninth Workshop on Statistical Machine Translation, Association for Computational Linguistics (2014)
5. Costa-Jussà, M.R., Farrús, M.: Statistical machine translation enhancements through linguistic levels: a survey. ACM Comput. Surv. **46**(3), Article 42 (2014)

6. Gale W.A., Church K.W.: A program for aligning sentences in bilingual corpora. Comput. Linguist. (1993)
7. Haghighi, A., Liang, P., Berg-Kirkpatrick, T., Klein, D.: Learning bilingual lexicons from monolingual corpora. ACL. Associat. Comput. Linguist. (2008)
8. Kay M., Röscheisen M.: Text-translation alignment. Computat. Linguist. (1993)
9. Koehn, P., Knight, K.: Learning a translation lexicon from monolingual corpora. In: Proceedings of ACL Workshop on Unsupervised Lexical Acquisition, ACL (2002)
10. Krajewski, R., Rybiński, H., Kozłowski, M.: A Seed Based Method for Dictionary Translation. Foundations of Intelligent Systems. LNAI. Springer, Berlin (2014)
11. Protaziuk, G., Wróblewska, A., Bembenik, R., Rybinski, H.: Lexical Ontology Layer—A Bridge between Text and Concepts. Foundations of Intelligent Systems. Springer, Berlin (2012)
12. Reinhard R.: Automatic identification of word translations from unrelated English and german corpora. In: Proceedings of the 37th annual meeting of the Association for Computational Linguistics on Computational Linguistics, ACL (1999)
13. Smadja, F., McKeown, K.R., Hatzivassiloglou, V.: Translating collocations for bilingual lexicons: a statistical approach. Comput. Linguist. (1996)
14. Wu D.: Toward machine translation with statistics and syntax and semantics, In: Proceedings of the IEEE Workshop on Automatic Speech Recognition & Understanding (ASRU'09) (2009)

VTLN Using Different Warping Functions for Template Matching

Maulik C. Madhavi, Shubham Sharma and Hemant A. Patil

Abstract In most automatic speech recognition (ASR) systems, speaker differences are compensated by *normalizing* the vocal tract lengths of the speakers. This is implemented by warping the frequency-axis by appropriate *warping* factor. However, it is computationally expensive to find warping factor for each speaker. This problem is overcome by incorporating a *universal* warping function for all the speakers. Different psychoacoustic scales have been proposed over the past decade that are assumed to be similar to the frequency response of *basilar membrane* (BM) of human auditory system. In this paper, different warping functions are studied with an aim of vocal tract length normalization (VTLN) and template matching experiments are done using dynamic time warping (DTW) algorithm to test the performance of various warping functions. It was observed that features obtained by warping the frequency-axis by psychoacoustic scales improve the classification performance. In particular, Equivalent Rectangular Bandwidth (ERB)-scale based warping improves the precision by 7.17 % over state-of-the-art mel frequency cepstral coefficients (MFCC) for template matching done on isolated digits of TIDIGITS database and 6.16 % on words from TIMIT database.

Keywords Vocal tract length normalization · Frequency warping · Dynamic time warping · Template matching

M.C. Madhavi (✉) · H.A. Patil
Dhirubhai Ambani Institute of Information and Communication Technology, Gandhinagar, India
e-mail: maulik_madhavi@daiict.ac.in
URL: http://www.daiict.ac.in

H.A. Patil
e-mail: hemant_patil@daiict.ac.in

S. Sharma
Indian Institute of Science, Bangalore, India
e-mail: shubham@mile.ee.iisc.ernet.in
URL: http://www.iisc.ernet.in

D. Ryżko et al. (eds.), *Machine Intelligence and Big Data in Industry*, Studies in Big Data 19, DOI 10.1007/978-3-319-30315-4_10

1 Introduction

Template matching is a technique of matching two isolated words with the help of acoustic features. In this paper, template matching is done using dynamic time warping (DTW) algorithm. In applications like keyword spotting, audio search, etc. the database of audio files may contain speech by different speakers (such as males, females or children). In addition, the search term or query speech may also be spoken by a different speaker. Even though the variations in acoustic properties in different speakers occur on account of various reasons, it is widely accepted that the major contribution in the variations is by differences in the lengths of vocal tract among speakers [1]. Vocal tract length normalization (VTLN) aims to compensate for the differences in lengths of vocal tracts of different speakers by warping the frequency-axis. Vocal tract length (VTL) varies from about 13 cm in adult female to 17.5 cm for adult male. The formant frequencies (i.e., $F_1, F_2, F_3, \ldots$) of female speakers are said to be about 20 % higher than that of male speakers [2]. By modeling the vocal tract as a uniform tube, formant frequencies are related to VTL by the following relationship [3]:

$$F_n = \frac{(2n-1)v}{4L}, \quad n \in \mathbb{Z}^+, \tag{1}$$

where v = velocity of sound ($\approx$345 m/s), L = VTL. For two speakers, say, speaker A and speaker B, with different VTLs (say L_A and L_B), their respective spectra are scaled versions of each other, i.e., $F_A(\omega) = F_B(\alpha_{AB}\omega)$, where the scaling factor α_{AB} is the ratio of the VTLs, i.e., $\alpha_{AB} = L_A/L_B$. Hence, speaker differences can be minimized by normalizing the spectra of the speakers by a scaling factor dependent on speakers. Warping functions other than the linear function have been proposed by various researchers. In particular, Eide and Gish proposed an empirical nonlinear warping function derived from the third formant frequency of the speakers [4]. Lee and Rose proposed piecewise linear warping function in [5]. In addition, scales obtained by experimental results on perception of pitch and loudness summation such as mel scale [6] and Bark scale [7], respectively have also been used for improving the performance of automatic speech recognition (ASR) systems. In this paper, performance of warping functions based on these psychoacoustic scales is compared with that of mel frequency cepstral coefficients (MFCC) for template matching of words and digits spoken by different speakers.

Organization of the remaining paper is as follows. Section 2 discusses various frequency warping functions used. Details of discrete implementation of warping are provided in Sect. 3. Feature extraction procedure is described in Sect. 4. Experimental results on TIMIT and TIDIGIT databases are discussed in Sect. 5. Finally, Sect. 6 summarizes the paper along with future research directions.

2 Frequency Warping Functions

Details of warping functions obtained by log scale, piecewise log scale, mel scale, Bark scale and ERB scale are discussed below:

2.1 Log Warping

Consider two speakers, *viz.*, speaker A and speaker B. Thus, their spectra are related as:

$$F_A(\omega) = F_A(\alpha_{AB}\omega) \tag{2}$$

Now, replacing ω by e^f for log-warping [8], we get,

$$F_A(e^f) = F_B(\alpha_{AB}e^f) = F_B(e^{(f+log\alpha_{AB})}), \Rightarrow f_A(f) = f_B(f + log\alpha_{AB}). \tag{3}$$

Hence, in the log-warped domain, the speaker-dependent scale factor (i.e., α_{AB}) becomes a *shift factor* [8]. Utilizing the shift-invariance property of Fourier transform (FT), VTL normalized features can be obtained. Taking Fourier transform (FT) of (3) with respect to f, we get,

$$Z_A(\xi) = FTf_A(f) = FTf_B(f + log\alpha_{AB}) = e^{j\xi log\alpha}Z_B(\xi), \tag{4}$$

$$\Rightarrow |Z_A(\xi)| = |e^{j\xi log\alpha}Z_B(\xi)| = |Z_B(\xi)| \tag{5}$$

Thus, the speaker-dependent factor appears in the phase part and can be removed by taking the magnitude as expressed in (4).

2.2 Piecewise Log-Warping

It is observed that the speaker-dependent warping factor α_{AB} is a function of frequency [8], i.e., $F_A(\omega) = F_B(\alpha_{AB}(\omega)\omega)$. Hence, use of piecewise log-warping is proposed. The frequency-axis is divided into five logarithmically equal parts. For TIMIT database with sampling frequency, $fs = 16{,}000$ Hz, the frequency bands in the range 100–7000 Hz are empirically taken as [100, 240) Hz, [240, 550) Hz, [550, 1280) Hz, [1280, 3000) Hz and [3000, 7000) Hz [8].

2.3 Mel Scale Warping

The scale of pitch, mel scale, was proposed by Stevens and Volkmann in 1940 [6]. They conducted experiments in which the observers were asked to divide three overlapping frequency ranges into four intervals of equal pitch distances. This scale is in great agreement with the function relating the frequency stimuli and corresponding distance on the basilar membrane (BM) from the experiments conducted on guinea pigs [6]. Although it is not necessary for an auditory-based scale to be useful for an ASR task, it is found that warping by mel scale outperforms the linear warping for VTLN [9]. Frequency in mels(f_{mel}) can be related with frequency in Hz(f) as [10]:

$$f_{mel} = 2595\, log_{10}\left(1 + \frac{f_{Hz}}{700}\right). \tag{6}$$

2.4 Bark Scale Warping

Bark scale is developed based on the concept of critical bands. Zwicker, Flottorp and Stevens in 1956 conducted some experiments on complex tones and noise to analyze the effect of critical bands on loudness summation [11]. It was observed that the loudness remains constant up to a certain critical frequency after which the loudness increases with increase in frequency. This band of frequency in which loudness is constant is known as the critical band. Zwicker in 1961 provided the relation between *critical band* function and frequency [7]. Subdivision of frequency scale into critical bands depicts high correlation with the filterbank mechanism of cochlea in human ear. One bark is said to be equal to the bandwidth of one critical band. Value of Bark scale varies up to 24 corresponding to first 24 filterbanks. Frequency in Bark (fBark) can be obtained for corresponding frequency in Hertz from the following expression [12]:

$$f_{Bark} = 13\, tan^{-1}\,(0.00076\, f_{Hz}) + 3.5\, tan^{-1}\left(\frac{f_{Hz}}{7500}\right)^2. \tag{7}$$

2.5 ERB Warping

Critical bandwidth is almost constant for centre frequencies below 500 Hz. However, it was observed that the *equivalent rectangular bandwidth* (ERB) of the critical bands decreases below 500 Hz [13]. ERB scale incorporates this variation and is given by the analytic expression [13]:

$$f_{ERB} = 11.17\, ln\left|\frac{f_{Hz} + 0.312}{f_{Hz} + 14.675}\right| + 43. \tag{8}$$

3 Discrete Implementation of Warping Functions

To implement non-uniform discrete Fourier transform (DFT), amplitude of Fourier transform is calculated at non-uniformly spaced points determined by the warping function used. Implementation is better explained by a schematic diagram shown in Fig. 1. Frequency in the warped-domain is divided into equal parts and is mapped back in the uniform frequency-domain (frequency in Hz). It is at these non-uniformly spaced points that the DFT is calculated. To find N-point non-uniform DFT, the warped frequency is divided into N equal points. For example, in case of log-warping, the nonuniform points are at a distance of

$$\Delta v = \frac{log\left(\frac{fs}{2}\right) - log(1)}{N} = \frac{log\left(\frac{fs}{2}\right)}{N}, \tag{9}$$

where fs is the sampling frequency. Now, log-warped DFT can be calculated as:

$$X(k) = \Sigma_{n=0}^{N-1}\left[x(n)exp(-j2\pi nT_s exp(k\Delta v))\right], \quad k = 0, 1, \ldots, N-1. \tag{10}$$

Similarly, other warping functions can be implemented using the following equation:

$$X(k) = \Sigma_{n=0}^{N-1}\left[x(n)exp(-j2\pi nT_s f_w^{-1}(k\Delta v))\right], \quad k = 0, 1, \ldots, N-1. \tag{11}$$

where f_w^{-1} is the inverse mapping to the frequency in Hz and f_w can be f_{mel}, f_{Bark} or f_{ERB}. Piecewise log-warping is incorporated by taking different number of points in each of the five frequency bands. Here, we have taken 34, 47, 84, 141 and 206 points in the five bands, respectively to make a total of 512 points. Figure 2 shows warping by log scale, piecewise log scale, mel scale, Bark scale and ERB scale in the frequency range 100–6999 Hz.

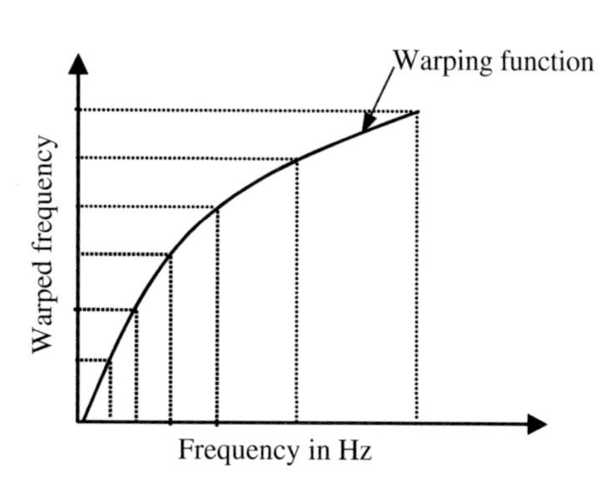

Fig. 1 A schematic representation to describe frequency warping

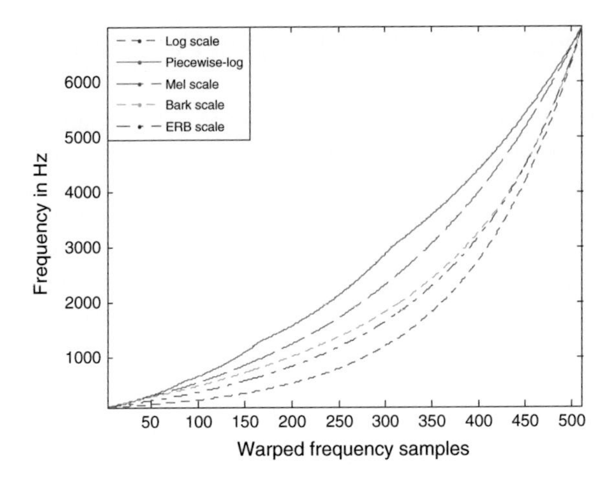

Fig. 2 Plots of different warping schemes

4 Feature Extraction

First the given speech signal is divided into frames of 25 ms duration with overlap of 10 ms. From the source-filter model of speech production, it can be said that speech signal is the output of the system (*viz.*, vocal tract filter) excited by the source (*viz.*, glottal flow). Even though the frame of 25 ms would be quasi-stationary, it still consists of pitch source harmonics. These pitch harmonics are suppressed by the following method. Each frame is divided into subframes of 6 ms with overlap of 4 ms and Hamming windowed. For each subframe, autocorrelation function is obtained. An average autocorrelation is calculated over all the subframes and multiplied with a Hamming window. Now, this averaged autocorrelation function is used to find the non-uniform DFT corresponding to different warping functions. This method essentially provides power spectral density (PSD) (by Wiener-Khinchin theorem) and the power spectrum does not contain any pitch harmonics. Finally, cepstral coefficients are obtained by taking discrete cosine transform (DCT) of logarithm of non-uniform DFT. Same procedure is applied for obtaining features warped by different warping functions. Figure 3 illustrates the detailed procedure of feature extraction.

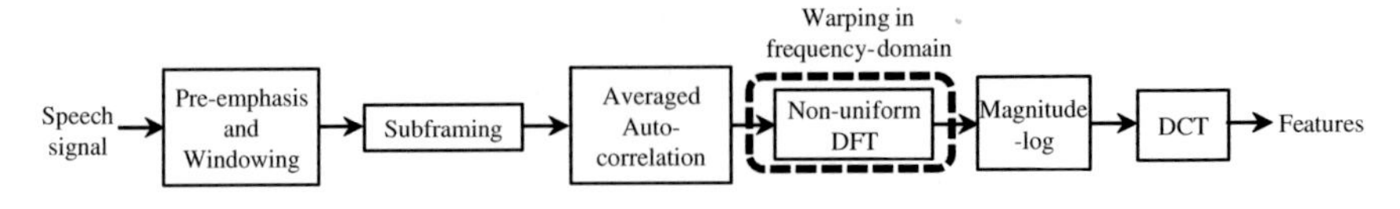

Fig. 3 Block diagram to illustrate procedure of feature extraction. After [14]

5 Experimental Results

Acoustic features obtained by different warping methods are used for template matching task. Each sample is taken as query and its distance is measured from all the other samples (i.e., reference) using Dynamic Time Warping (DTW) algorithm. DTW is a dynamic programming (DP) algorithm to match two sequences (i.e., utterances) of different lengths (i.e., durations). Best warping function is estimated using DTW to obtain the optimal alignment path for the two sequences [15]. In DTW, the time-axis is warped for appropriate matching. By intuition, the distance should be less between the samples of the same word and it should be higher when samples of different words are compared. The performance of warped features is compared with state-of-the-art mel frequency cepstral coefficients (MFCC). Here, 39-D feature vectors consisting of 13-static coefficients (C_1–C_{13}), 13-delta and 13-acceleration coefficients are considered. Experiments are performed on two different databases, *viz.*, TIMIT and TIDIGIT.

5.1 Experiment on TIMIT Database

The first set of experiments is conducted on 50 words, taken from the continuous speech of TIMIT database from different speakers [16]. The number of instances of each word is about 30–50 times in the database. Here, two performance measures, *viz.*, *% precision @ N* (N is the total number of occurrences on test) and *% equal error rate (EER)*, have been used to compare the performance of the feature sets. Precision is the ratio of correctly detected words to total number of words. % EER is computed as the point where false acceptance rate (FAR) is equal to the false rejection rate (FRR). Hence, lower the EER, better the feature set. Precision and EERs obtained from different feature sets (MFCCs and log-warped, piecewise log-warped, mel-warped, Bark warped and ERB warped features) are listed in Table 1. It can be observed that the precision is improved for warped features significantly. The detection error trade-off (DET) curve is shown in Fig. 4. DET curve shows the trade-off between false acceptance and false rejection events. The DET curve for warped features is below the DET curve of MFCC. This shows the better classification ability

Table 1 % Precision and % Equal error rates (EER) obtained by different feature sets (in %) for TIMIT database

Feature set warping	MFCC	Log warping	Piecewise log warping	Mel warping	Bark warping	ERB warping
Precision	36.69	42.80	36.89	38.67	41.88	42.85
EER	18.68	14.47	17.44	17.01	15.71	15.18

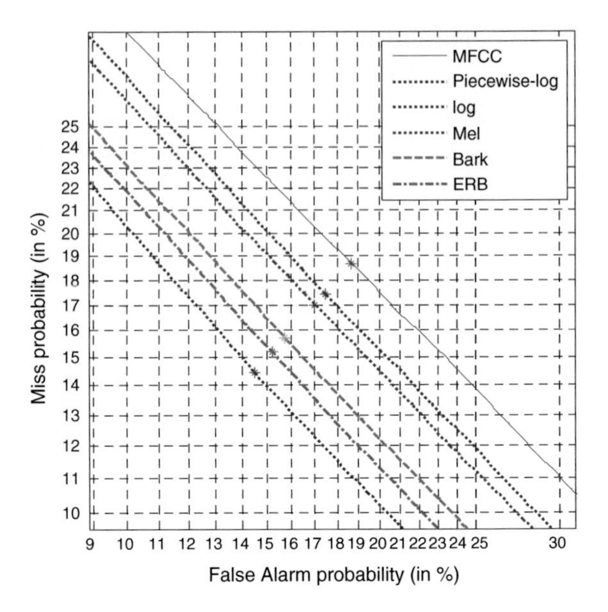

Fig. 4 DET curves for MFCC, log-warped, piecewise log-warped, mel-warped, Bark-warped and ERB-warped feature sets for words from TIMIT database

of warped feature sets using DTW algorithm. Warping with ERB scale shows highest % precision and log-warped features show minimum % EER.

5.2 *Experiment on TIDIGIT Database*

Another set of experiments is performed on ten isolated spoken digits (zero, one, two,..., nine) spoken by 20 speakers from TIDIGITS database [17] (i.e., 200 samples in total). TIDIGITS database consists of different combinations of digits spoken by adult men and women as well as girl and boy children. In this case, five speakers from each set (man, woman, boy and girl) are considered. From each speaker, each isolated spoken digit is taken and silence at the start and end of spoken digit is removed. Short-time energy is used to remove silence portions of speech data. Since there is no energy during fricatives /f/ and /s/ as in /four/, /five/, /six/ and /seven/, major regions of these fricatives are cut-off leading to distortions in the speech signals. Samples with high distortion are manually replaced. Similar to the previous experiment on isolated words of TIMIT database, 39-D MFCC, warped features with different warping functions are extracted for all the samples and distances between each sample is calculated using DTW algorithm. % Precision and % EERs obtained for different feature sets are provided in Table 2. It can be observed that the better performance of warped features is consistent, i.e., their ability to detect the similar words is higher than that of MFCCs even for a test set having higher speaker variations. It can be observed that there is a significant improvement in precision for warped features using ERB-warping as compared to other kinds of warping functions under study.

Table 2 % Precision and % Equal error rates (EER) obtained by different feature sets (in %) for TIDIGIT database

Feature set	MFCC	Log warping	Piecewise log warping	Mel warping	Bark warping	ERB warping
Precision	49.48	54.46	55.59	52.38	56.53	56.65
EER	24.59	21.33	19.57	21.11	19.19	19.11

However, the performance of Bark warped features is very similar to ERB scale warped features. Figure 5 shows the DET curves for all the feature sets under study. DET curves for features obtained by warping based on ERB scale and Bark scale are the lowest in the entire range. Hence, their performance is better than other warping scales. The main difference between Bark scale and mel scale is that the Bark scale has *critical bands* as the natural scale unit whereas mel scale maps the frequency to *distance along the BM*. The observations suggest that in this case, filterbank model of cochlea provides better discrimination of words. In addition, the similarity in the performance of Bark scale and ERB scale can be understood from the fact that ERB scale is proportional to the Bark scale for frequencies higher than 500 Hz. However, ERB scale decreases for frequencies below 500 Hz. Hence, ERB scale can be said to be a better representation of the filterbank model of the cochlea [12]. This could be a justification for better performance of features obtained by ERB scale-based warping as compared to other psychoacoustic scales. However, further investigation is desired in this area.

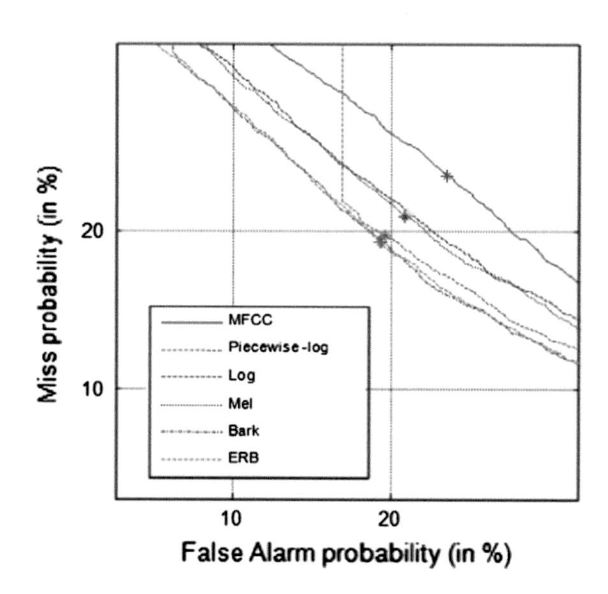

Fig. 5 DET curves for MFCC, log-warped, piecewise log-warped, mel-warped, Bark-warped and ERB-warped feature sets for words from TIDIGIT database

6 Summary and Conclusions

In this paper, template matching experiments were performed for isolated words of TIMIT database and digits of TIDIGITS database. Different warping functions were discussed and their performances were compared. It is observed that although the warping functions are similar, there is difference in their recognition abilities. Warping functions based on psychoacoustic scales perform better than state-of-the-art MFCC features. Out of all the warping functions considered in this study, ERB scale-based warping is found to perform better than all others as it provides better classification precision and least % EER for both the experiments. It can be concluded that warping based on scales that mimic the auditory mechanism of speech perception of human beings indeed improve the classification accuracy and are useful for recognition purposes including speakers of different genders. Detailed analysis of different warping functions needs to be done and is aimed in future.

Acknowledgments The authors would like to thank Department of Electronics and Information Technology (DeitY), Govt. of India for sponsoring the consortium project, "Development of Prosodically Guided Phonetic Engine for Searching Speech Databases in Indian Languages". They also thank the consortium leader Prof. B. Yegnanarayana (IIIT Hyderabad) and the authorities of DA-IICT to support this research work.

References

1. Fant, G.: Acoustic Theory of Speech Production. Mouton & Co, The Hague (1960)
2. Fant, G.: A note on vocal tract size factors and non-uniform f-pattern scalings. Speech Transm. Lab. Q. Progress Status Rep. **1**, 22–30 (1966)
3. Quatieri, T.F.: Discrete-time Speech Signal Processing: Principles and Practice. Pearson Education, London (2006)
4. Eide, E., Gish, H.: A parametric approach to vocal tract length normalization. In: International Conference on Acoustic Speech & Signal Processing, ICASSP, vol. 1, pp. 346–348. IEEE, Atlanta, Georgia, USA (1996)
5. Lee, L., Rose, R.: A frequency warping approach to speaker normalization. IEEE Trans. Speech Audio Process. **6**(1), 49–60 (1998)
6. Stevens, S.S., Volkmann, J.: The relation of pitch to frequency: A revised scale. Am. J. Psychol. **53**, 329–353 (1940)
7. Zwicker, E.: Subdivision of the audible frequency range into critical bands (frequenzgruppen). J. Acoust. Soc. Am. **33**(2), 248 (1961)
8. Umesh, S., Cohen, L., Marinovic, N., Nelson, D.: Frequency-warping in speech. In: 4th International Conference on Spoken Language Processing, ICSLP, vol. 1, pp. 414–417. IEEE (1996)
9. Kamm, T., Hermansky, H., Andreou, A.G.: Learning the Mel-scale and optimal VTN mapping. In: Center for Language and Speech Processing, Workshop (WS 1997). Johns Hopkins University (1997)
10. O'shaughnessy, D.: Speech Communication: Human and Machine. Universities press (1987)
11. Zwicker, E., Flottorp, G., Stevens, S.S.: Critical band width in loudness summation. J. Acoust. Soc. Am. **29**(5), 548–557 (1957)
12. Zwicker, E., Terhardt, E.: Analytical expressions for critical-band rate and critical bandwidth as a function of frequency. J. Acoust. Soc. Am. **68**(5), 1523–1525 (1980)

13. Moore, B.C.J., Glasberg, B.R.: Suggested formulae for calculating auditory-filter bandwidths and excitation patterns. J. Acoust. Soc. Am. **74**(3), 750–753 (1983)
14. Umesh, S., Sanand, D.R., Praveen, G.: Speaker-invariant features for automatic speech recognition. In: IJCAI, pp. 1738–1743 (2007)
15. Sakoe, H., Chiba, S.: Dynamic programming algorithm optimization for spoken word recognition. IEEE Trans. Audio Speech Lang. Process. **26**(1), 43–49 (1978)
16. Garofolo, J.S., Lamel, L.F., Fisher, W.M., Fiscus, J.G., Pallett, D.S., Dahlgren, N.L.: DARPA TIMIT acoustic-phonetic continous speech corpus CD-ROM (1993)
17. Leonard, R.G., Doddington, G.: Tidigits ldc93s10. Linguistic Data Consortium, Philadelphia (1993)

A Comparative Study on Music Genre Classification Algorithms

Wojciech Stokowiec

Abstract Music Genre Classification is one of the fundamental tasks in the field of Music Information Retrieval (MIR). In this paper the performance of various music genre classification algorithms including Random Forests, Multi-class Support Vector Machines and Deep Belief Networks is being compared. The study is based on the "Million Song Dataset" a freely-available collection of audio features and metadata. The emphasis is put not only on classification accuracy but also on robustness and scalability of algorithms.

Keywords Music genre recognition · Million Song Dataset · Machine learning

1 Introduction

Musical genres are categorical descriptions that are used to characterize music. Although classification criteria may seem subjective and arbitrary, humans have shown remarkable skill at genre recognition. As the number of songs, which are available to listeners, grows exponentially, it seems implausible that they would be equipped with appropriate textual information such as music genre. Music genre classification is considered to be a great practical component of music retrieval and recommendation systems, thus techniques for automatic genre classification would be a valuable tool.

This work describes systems of automatic music genre recognition based exclusively on audio features. The paper is structured as follows: firstly, a brief review of related work is presented in Sect. 2. Secondly, the dataset, on which this study has been based, is detailed in Sect. 3. Next, the algorithms that have been compared are described in Sect. 4 and results are given in Sect. 5. Finally, conclusions and future work can be found in Sect. 6.

W. Stokowiec (✉)
National Information Processing Institute, Warsaw, Poland
e-mail: wojciech.stokowiec@opi.org.pl

D. Ryżko et al. (eds.), *Machine Intelligence and Big Data in Industry*,
Studies in Big Data 19, DOI 10.1007/978-3-319-30315-4_11

2 Related Work

Although Music Genre Classification is a vibrant research field, it is difficult to find papers based directly on Million Song Dataset Benchmarks (MSDB).

Nevertheless, several classification models have been trained on features provided by MSD. For instance, [4] derived beat-aligned timbre and chroma features from music audio data contained in MSD in an effort to train a convolutional Deep Belief Network on all the data, and then used the computed parameters to initialize a convolutional multilayer perceptron. It achieved 29.5 % accuracy on genre recognition task with 20 genres and concluded that the gains obtained with pretraining are rather modest and are not advantageous for genre classification. Another interesting study has been conducted by [7], where authors proposed a framework of model blending based on combing features from audio and lyrics, which lead to accuracy of 38.6 % (10 hand-picked genres).

Several articles that exploit Deep Belief Networks in the context of automatic music genre classification have been published. In [14], DBN have been trained on GTZAN [12] dataset using greedy, layer-wise, unsupervised learning algorithm with short and long-term features, additionally fine-tuned via back propagation algorithm. They were able to reach 78.7 % accuracy, whereas best-performing, widely used classifier (SVM) achieved 75.9 % on the same dataset. It is worth noticing, that in GZTAN dataset there are only 10 classes, which renders the classification task easier. For comparison, our highest scoring algorithm (Random Trees) trained on dataset with 13 classes reached 62 % accuracy. One can claim, that in the setting of 10 genres classification task the performance of Random Trees would appropriately increase.

To our knowledge, our study is the first to evaluate Deep Belief Networks on MSDB dataset in the context of music genre recognition.

Authors of MSDB conducted a preliminary analysis of the classification algorithms such as Naive Bayes, Support Vector Machines, k-nearest Neighbours ($k = 1$), J48 Decision Tree and Random Forest using WEKA Machine Learning Toolkit [10]. They received best results on Statistical Spectrum Descriptors (with 168 dimensions) feature dataset with MSD Allmusic Genre Dataset (21 classes) with 66 % training set split. Highest accuracy has been achieved with Support Vector Machines and k-NN classifiers, both of which scored more than 27 % on SSD feature set. This is in contrast to our results, which suggest that Random Forest and Decision Trees are top performers.

A comprehensive study of recent advances in Music Genre Recognition can be found in [11].

3 Dataset

Million Song Dataset (MSD) is a freely-available collection of audio features and metadata for a million contemporary popular music tracks. It contains approximately 280 GB of data with 1,000,000 song files and 44,745 unique artists [2]. Traditionally, due to licensing issues, research in music information retrieval on commercial-scale dataset was limited to the industry. However, this has been circumvented by providing precomputed features instead of raw audio. The MSD enables researchers to test algorithms on a large-scale collection in real-world-like environments [10]. Unfortunately, some authors [9] claim that the absence of accurate documentation of the extraction algorithm renders the audio features provided by MSD of limited use.

In this study we used Million Song Dataset Benchmarks (MSDB) where wide range of audio features have been extracted from audio samples downloaded from external content provider. Detailed description of extraction procedure and employed software can be found in [10]. An overview of features is given in Table 1.

There are three datasets mapping individual tracks to their appropriate, expert annotated genre. The first one, the MSD Allmusic Genre Dataset (MAGD) consist of 21 genres. The second one, MSD Allmusic Top Genre Dataset (top-MAGD), consist of 13 genres—the top 10 genres from MAGD including three additional ones (Vocal, Folk and New Age). The last one, the MSD Allmusic Style Dataset (MASD), has 25 classes. All classes from aforementioned datasets are listed in Tables 2 and 3.

For detailed description of data collection and dataset building please refer to [10].

In order to facilitate repeatability of experiments, several partitions of the dataset have been prepared in the MSDB. We have decided to use stratified partition with frequently used 2/3 training and 1/3 test split and partition with fixed number of training samples, equally sized for each class with 2,000 samples per genre.

Table 1 Overview of selected features from Million Song Dataset Benchmarks

#	Feature set	Dimensions
1	Rhythm patterns	1440
2	Rhythm histograms	60
3	Temporal rhythm histograms	420
4	MARSYAS timbral features	124
5	MFCC features	26

Table 2 MSD Allmusic Genre Dataset (MAGD)—upper part represents the MSD Allmusic Top Genre Dataset (Top-MAGD)

Genre name	Number of tracks
Pop/rock	238,786
Electronic	41,075
Rap	20,939
Jazz	17,836
Latin	17,590
R&B	14,335
International	14,242
Country	11,772
Religious	8,814
Reggae	6,946
Blues	6,836
Vocal	6,195
Folk	5,865
New age	4,010
Comedy/spoken	2,067
Stage	1,614
Easy listening	1,545
Avant-garde	1,014
Classical	556
Childrens	477
Holiday	200
Total	**422,714**

4 Algorithms

In this study several classification algorithms have been compared: Decision Tree (DT), Random Forest (RF), Multi-class Support Vector Machine (SVM), Multinomial Logistic Regression (LR) and Deep Belief Network (DBN). Short description of selected algorithms are given in the following subsections. As far as algorithm's implementations are concerned, Apache Spark's MLlib has been used for training Decision Trees, Random Forest, and binary classifiers such, as SVM and Logistic Regression. Due to the fact, that Multi-class classification is currently not supported for SVM and Logistic Regression, we have chosen to implement them in Scala, building upon MLlib binary classificators. Deep Belief Networks have been trained using DL4J, an open-source library written for Java and Scala.

Table 3 MSD Allmuisc Style Dataset (MASD)

Genre name	Number of tracks
Pop indie	18,138
Rock college	16,575
Rock contemporary	16,530
Hip hop rap	16,100
Dance	15,114
Metal alternative	14,009
Pop contemporary	13,624
Rock hard	13,276
Rock alternative	12,717
Experimental	12,139
Country traditional	11,164
Rock neo psychedelia	11,057
Electronica	10,987
Metal heavy	10,784
Jazz classic	10,024
Metal death	9,851
Folk international	9,849
Punk	9,610
Pop latin	7,699
Gospel	6,974
Blues contemporary	6,874
Grunge emo	6,256
RnB soul	6,238
Reggae	5,232
Big band	3,115
Total	**273,936**

4.1 Decision Tree

Decision tree is a greedy classification algorithm that performs a recursive binary split of the feature space [5]. In our study Gini impurity measure has been used as a splitting criterion. Our preliminary study has shown that it yields best results on Million Song Dataset. Moreover, based on empirical considerations (grid-search with 3-fold cross-validation), the maximum depth of a tree has been limited to 10 and the number of bins used when discretizing continuous features has been set to 64.

4.2 Random Forest

Random forests are ensembles of Decision Trees [5]. They are especially appealing in the context of big-data because of the fact, that individual trees can be built independently, thus learning them is inherently parallel. Since trees should be built on subset of data, it straightforward to parallelize the whole process. As it was the case with Decision Trees, the hyperparameters have been optimised using grid-search in conjunction with 3-fold cross-validation.

4.3 Multi-class SVM

The Support Vector Machine (SVM) is a state-of-the-art linear binary classification algorithm. In this study we have decided to take advantage of the fact, that (in opposition to SVMs employing kernel-trick) linear SVM scale well with the number of examples [3, 5]. Moreover, the naive way of explicitly computing non-linear features does not scale well with the number of input features and in the case of Million Song Dataset occurred to be computationally prohibitive.

We have decided to employ One-Vs-The-Rest strategy to construct Multi-class SVM. It is worth noticing, that in the case of a multi-label classification task, choosing the category based on maximal posterior probability over all classes is the Bayes optimal decision for the equal loss case [8]. Unfortunately, Standard SVM do not directly provide posterior probability estimates, therefore those parameters have to be manually calibrated. Based on the work [8] we have decided to use a parametric model to fit the posterior probabilities directly. Following parametrization has been used:

$$P(y = 1 \mid x) \approx P_{A,B}(f) = \frac{1}{1 + \exp(Af + B)}, \quad \text{where } f = f(x). \tag{1}$$

The parameters A and B from Eq. 1 are fit by solving following maximum likelihood problem from a training set $(f_i,\ y_i)$:

$$\arg \min_{A,B \in \mathbb{R}} F(z) = -\sum_{i=1}^{n} \Big(y_i \log(p_i) + (1 - y_i) \log(1 - p_i) \Big), \tag{2}$$

where $p_i = P_{A,B}(f_i)$ and n is the number of training examples.

According to [8] for linear SVMs the bias introduced by using the same data set for training the model and estimating parameters from Eq. 1 is negligible for large datasets. This is due to the fact, that as in almost all cases, a maximum of $N + 1$ support vectors will lie on the margin (were N is the dimensionality of the input vector). In our case of linear SVM no additional preprocessing is needed.

4.4 Deep Belief Network

For classification, a DBN with ℓ layers models the joint distribution between target y, observed variables x_j and i hidden layers $\mathbf{h}^k$ made of all binary units h_i^k, as follows [6]:

$$P(\mathbf{x}, \mathbf{h}^1, \ldots, \mathbf{h}^\ell, y) = \left(\prod_{k=1}^{\ell-2} P(\mathbf{h}^k | \mathbf{h}^{k+1}) \right) P(y, \mathbf{h}^{\ell-1}, \mathbf{h}^{\ell-1}), \tag{3}$$

where $\mathbf{x} = \mathbf{h}^0$, $P(\mathbf{h}^k|\mathbf{h}^{k+1})$ is a conditional distribution for the visible units conditioned on the hidden units of the Restricted Boltzmann Machine (RBM) at level k and $P(y, \mathbf{h}^{\ell-1}, \mathbf{h}^{\ell-1})$ is the visible-hidden joint distribution in the top-level RBM. A RBM has the following form:

$$P(\mathbf{x}, \mathbf{h}) \propto \exp(\mathbf{h}' W \mathbf{x} + b' \mathbf{x} + c' \mathbf{h}) \tag{4}$$

with parameters $\theta = (W, b, c)$. During network training, contrastive divergence has been run only once, which has been shown to work surprisingly well. In an effort to adapt RBM to accept continuous input we have employed a Gaussian transformation on the visible layer and a rectified-linear-unit transformation on the hidden layer. Initial weights has been sampled from uniform distribution. We have decided to use different architectures depending on feature dataset. Number of hidden layers and their size has been chosen after initial empirical investigation.

5 Results

Our experimental results show that Random Forests and Decision Trees outperform Multi-class SVM, Multinomial Logistic Regression and naively trained DBN independently of the chosen features, genre dataset or benchmarking partitions. Highest scoring algorithm (Random Forest) trained on dataset with 13 classes reached accuracy of 62 %.

Random Forests and Decision Trees are also easily parallelizable, which effectively makes them the algorithm of choice in setting with time-limited computing resources. In the case of DBN hyperparameter tuning, using grid-search has occurred to be computationally intractable and time constraints become an important issue. All of this led to unsatisfactory results.

In all cases Multi-class SVM had at least 10 % of accuracy less than Random Forest. One possible remedy for SVM's poor performance may be changing the way multiple classes are handled. It is worth investigating whether employing One-Vs-One strategy, instead of One-Vs-The-Rest, can bring benefits. We have conducted preliminary study in which 78, i.e. all possible genre pairs in top-MAGD dataset, One-Vs-One Random Forests have been trained. Results are promising: mean F-1 measure was equal to 81 and 90 % of scores were above 68 %. Highest and lowest

Table 4 5 highest One-Vs-One classfier F-1 scores for top-MAGD genres on Rhythm Histograms dataset

Genre	F-1 score
New age vs Pop rock	0.9834
Pop rock vs folk	0.9761
Vocal vs Pop rock	0.9746
Blues vs Pop rock	0.9721
Pop rock vs Reggae	0.9719

Table 5 5 lowest One-Vs-One classfier F-1 scores for top-MAGD genres on Rhythm Histograms dataset

Genre	F-1 score
Blues vs Folk	0.6351
International vs RnB	0.6313
International vs Latin	0.6055
RnB vs Latin	0.5914
Vocal vs Folk	0.5818

scoring pairs are shown in Tables 4 and 5 respectively. It is also worth noticing, that SVM's performance decreased with the increase in the size of the input vector: SVM reached highest F-1 scores on MFCC features dataset (26 features) and lowest on Temporal Rhythm Histograms feature dataset (420 features).

Interestingly, audio features perform quite poorly on guitar based styles from MSD Allmusic Style Dataset (MASD). Quite unexpectedly, classifiers have problems distinguishing between *Rock College* and *Metal Heavy*. Similar observation have been made by [7].

Because of the space considerations we have decided to restrict result presentation to Temporal Rhythm Histograms, Rhythm Histograms, MFCC features and MARSYAS timbral features datasets with 66 % training, stratified set split. Additionally, we have included random classifier choosing each class with equal probability, denoted R, as our point of reference. Results are shown in Tables 6, 7, 8 and 9.

Table 6 Classification F-1 score for genre datasets on Temporal Rhythm Histograms feature dataset

Genre	DT	RF	SVM	LR	DBN	R
MAGD	0.59	0.60	0.12	0.11	0.05	0.048
top-MAGD	0.61	0.62	0.20	0.25	0.09	0.077
MASD	0.15	0.17	0.06	0.07	0.05	0.040

Table 7 Classification F-1 score for genre datasets on Rhythm Histograms dataset

Genre	DT	RF	SVM	LR	DBN	R
MAGD	0.58	0.59	0.37	0.35	0.09	0.048
top-MAGD	0.60	0.61	0.38	0.36	0.11	0.077
MASD	0.16	0.18	0.10	0.09	0.05	0.040

Table 8 Classification F-1 score for genre datasets on MFCC features dataset

Genre	DT	RF	SVM	LR	DBN	R
MAGD	0.58	0.59	0.49	0.48	0.05	0.048
top-MAGD	0.61	0.62	0.51	0.50	0.08	0.077
MASD	0.18	0.21	0.16	0.15	0.05	0.040

Table 9 Classification F-1 score for genre datasets on MARSYAS timbral features dataset

Genre	DT	RF	SVM	LR	DBN	R
MAGD	0.61	0.62	0.52	0.53	0.11	0.048
top-MAGD	0.64	0.65	0.55	0.54	0.10	0.077
MASD	0.21	0.22	0.17	0.18	0.05	0.040

6 Conclusions and Future Work

Although Random Forest and Decision Trees yield satisfactory results, several things can be done to improve the classifier performance. To be more explicit, the following three aspects are worth studying:

1. Inclusion of text features;
2. Exploration of different pairwise coupling methods in an effort to obtain better probability estimates for Multi-class classification;
3. DBN adjustment and optimization;

Ad (1) At the intuitive level, the inclusion of lyrical features can boost classifier accuracy, as lyrics cover semantic information about song's contents not available in the audio features. Additionally, in many music sub-genres the dividing line is often subtle and runs through the topics discussed rather than artistic means of expression. Encouraging results have been obtained in [7], where authors achieved 40 % accuracy based solely on bag-of-words lyric features. One can suggest, that, to a certain extent, lyrics and audio features are orthogonal and combining them in single model can yield better accuracy. It is also worth investigating whether using Markov Models or TF-IDF can bring additional benefits.

Ad (2) Platt [8] uses a Levenberg–Marquardt algorithm to solve Eq. 2. Instead other methods for solving unconstrained optimization can be used in an effort to boost classifier performance. Furthermore, as show in [13] pairwise coupling in conjunction with Platt scaling can yield satisfactory results. It is worth studying whether employing different probability estimation methods can boost classifier accuracy.

Ad (3) In future work, we would like to refine couple of aspect concerning the architecture of the network, such as the number and size of the hidden layers. Our experience with deep architectures shows that hyper-parameter optimization in large and multilayer models is not by any means an easy task. Recently, [1] showed that two sequential model-based optimization algorithms could outperform domain experts in the tuning Deep Belief Networks. It seems that it would be beneficial to incorporate algorithms proposed by [1] in subsequent works involving DBN.

Moreover, one can experiment with alternative ways of examining music genre classification system performance, for example, using multi-label classification may provide additional insight.

References

1. Bergstra, J., Bardenet R., Bengio Y., Kegl, B.: Algorithms for hyper-parameter optimization. In: Proceedings of the 24th Neural Information Processing Systems (NIPS 2011) (2011)
2. Bertin-Mahieux, T., Ellis, D., Whitman B., Lamere P.: The million song dataset. In: Proceedings of the 12th International Conference on Music Information Retrieval (2011)
3. Bishop, C.M.: Pattern Recognition and Machine Learning. Springer, Heidelberg (2007)
4. Dieleman, S., Brakel, P., Schrauwen, B.: Audio-based music classification with a pretrained convolutional network. In: Proceedings of the 12th International Society for Music Information Retrieval Conference (2011)
5. Hastie, T., Tibshirani, R., Friedman, J.H.: The Elements of Statistical Learning. Springer, New York (2001)
6. Hinton, G.E., Osindero, S., Teh, Y.: A fast learning algorithm for deep belief nets. Neural Comput. **18**, 1527–1554 (2006)
7. Liang, D., Gu, H., O'Connor, B.: Music genre classication with the million song dataset. Machine Learning Department, CMU (2011). http://www.ee.columbia.edu/~dliang/files/FINAL.pdf
8. Platt, J.: Probabilistic outputs for support vector machines and comparisons to regularized likelihood methods. Advances in Large Margin Classifiers, pp. 61–74. MIT Press, Cambridge (1999)
9. Schindler, A., Rauber, A.: Capturing the temporal domain in Echonest Features for improved classification effectiveness. In: Proceedings of the 10th International Workshop on Adaptive Multimedia Retrieval (2012)
10. Schindler, A., Mayer, R., Rauber, A.: Facilitating comprehensive benchmarking experiments on the million song sataset. In: Proceedings of the 13th International Society for Music Information Retrieval Conference (2012)
11. Strum, B.L.: A survey of evaluation in music genre recognition. Adaptive multimedia retrieval: semantics, context, and adaptation. Lect. Notes Comput. Sci. **8382**, 29–66 (2014)
12. Tzanetakis, G., Cook, P.: Musical genre classification of audiosignals. IEEE Trans. Audio Speech Process. **10**(5), 293–302 (2002)
13. Wu, T.F., Lin, C.J., Weng, R.C.: Probability estimates for multi-class classification by pairwise coupling. JMLR **5**, 975–100 (2004)
14. Yang, X., Chen, Q., Zhou, S., Wang, X.: Deep belief networks for automatic music genre classification, In: Proceedings of the 12th Annual Conference of the International Speech Communication Association (2011)

Part IV
Software

Information Selection and Data Compression RapidMiner Library

Marcin Blachnik and Mirosław Kordos

Abstract We present an Information Selection and Data Compression RapidMiner Library, which contains several known instance selection algorithms and several algorithms developed by us for classification and regression tasks. We present the motivation for creating the library and the need for developing new instance selection algorithms or extending the existing ones. We discuss how the library works and how to use it.

1 Introduction

A huge change has been observed during the last decade of computational intelligence applications. While most of the state of the art machine learning methods have remained unchanged, the volume of data has been continuously growing. This has lead to the problem of tractability of the training sets of such enormous sizes. In the past the most common problem was that the datasets were too small and in consequence the problem of over-fitting occurred. Nowadays we frequently have to deal with very big datasets (see Fig. 1), what leads to a very rapid growth of calculation time, especially that the computational complexity of many algorithms grow faster than the dataset size.

Figure 1 shows the average size of the datasets submitted to the UCI [1] repository per year. What is also interesting, the size of the training set often influences not only the training phase of building a prediction model but also the prediction phase. This applies not to all models as for example MLP neural networks have fixed internal

M. Blachnik (✉)
Department of Applied Informatics, Silesian University of Technology,
Krasińskiego 8, Katowice, Poland
e-mail: marcin.blachnik@polsl.pl

M. Kordos
Department of Computer Science and Engineering, University of Bielsko-Biala,
Willowa 2, Bielsko-biała, Poland
e-mail: mkordos@ath.bielsko.pl

D. Ryżko et al. (eds.), *Machine Intelligence and Big Data in Industry*,
Studies in Big Data 19, DOI 10.1007/978-3-319-30315-4_12

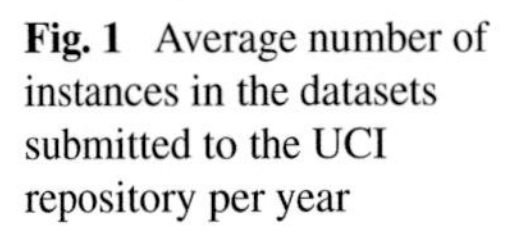

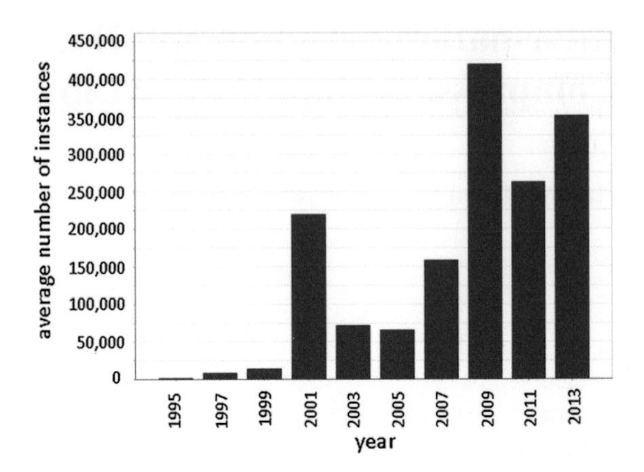

Fig. 1 Average number of instances in the datasets submitted to the UCI repository per year

representation but it influences for example the popular SVM models as they perform summation over support vectors, which are a subset of training examples.

As a result of these observations the question arise "Do we really need all these data to build an accurate model?" and in consequence the question if we can reduce the size of the data without seriously affecting the accuracy of the model. Results of many researchers showed that it is possible and even more, it is possible to build a more accurate model using only the selected data. For example see [8, 12]. As it is so important, we developed new instance selection algorithms to address several problems not dealt promptly so far and we have developed a library for instance selection and optimization for the well-known data mining software—RapidMiner. The library contains as well the known instance selection algorithms as our new ideas. We will briefly describe the architecture of this *Information Selection and Data Compression* library with key features and some examples. The paper is organized as follows: first all of the instance selection and prototype optimization methods are described, then in the next section the internal structure of the library is presented. The following section presents usage examples and the final section summarizes the paper.

For simplification we use the following notation $\mathbf{T}$—a dataset, which consists of n—instances, where a single instance i is represented as a tuple $\mathbf{T}_i = [\mathbf{x}, y]$ where $\mathbf{x} \in \Re^m$ is a vector in m dimensional space, and the value of y depends on the problem: $y \in [c_1, c_2, \ldots, c_l]$ for classification tasks or $y \in \Re$ for regression tasks.

2 The Algorithms

The goal of information selection methods is to construct a dataset, which fulfills the inequality $n > n'$, where n' is the size of the new dataset. In the literature two families of information selection methods can be found. The task of the first family is to select a subset of examples, where all the selected examples are the original instances from the training set. The second family, which is called prototype optimization methods, does not require the selected subset to consist of the original examples. The *Information Selection and Data Compression* library implements both of these families.

2.1 Instance Selection Algorithms

In the group of instance selection methods 12 state of the art algorithms are implemented. All of the implemented algorithms are presented in Table 1, but because of the limited size of the paper, the algorithms are not discussed in details, instead an appropriate reference is given. A general comparison of instance selection methods is provided in [8, 9].

As most of the instance selection methods were designed for classification tasks, we redesigned and extend them to support also regression problems. For this purpose the standard class label equality condition between the predicted and the true instance label was replaced by the inequality, which measures the error defined as a difference between the predicted value and the true label value. If it is greater than a certain threshold θ then the appropriate action is performed.

Table 1 List of supported instance selection algorithms

Algorithm	Description	Problem type		References
		Classification	Regression	
CNN	Condensed nearest neighbor rule	+	+	[10]
ENN	Edited nearest neighbor rule	+	+	[18]
Random	Selects instances by random	+	+	
RMHC	Random mutation hill climbing	+	+	[15]
All-kNN	All-kNN algorithm	+	+	[17]
MC	montecarlo based instance selection	+	+	[15]
RNG	Relative neighbor graph	+	+	[16]
RENN	Repeated ENN	+	+	[17]
GE	Gabriel editing	+	+	[16]
ELH	Encoding length heuristics	+	–	[7]
IB2	IB2 algorithm	+	–	[2]
IB3	IB3 algorithm	+	–	[2]

$$\text{if } (\|Y_{real} - \bar{Y}_{predicted}\| > \theta) \text{ then } do\ action. \tag{1}$$

It also supports an extended version of this loss type, where the difference is measured locally taking into account the local standard deviation restricted to k nearest neighbors of the query instance.

$$\text{if } (\|Y_{real} - \bar{Y}_{predicted}\| > \theta \cdot std(k)) \text{ then } do\ action \tag{2}$$

For details see [13]. The library also provides access to the Weka Instance Selection Library, which also implements the family of Drop algorithms, HMNE, ICF, MI, MSS and RNN algorithm.

The library also contains a set of generalized and ensemble instance selection methods. The group of generalized instance selection methods, which includes Generalized-ENN and Generalized-CNN, supports any kind of prediction models (not only the originally used k-NN), which can be wrapped and used to judge if an instance should be rejected or not. The second group—ensemble methods contains four algorithms. They are used to improve the quality of single instance selection algorithm by creating a committee of individuals. The following ensemble algorithms are implemented:

- **Bagging**—where any single instance selection algorithm is performed p times on a random subset of the training set **T** and then the final acceptance/rejection decision is made by voting.
- **Voting**—where instead of using the same algorithm on different subsets, the diversity is provided by voting of different instance selection algorithms performed on the same dataset.
- **Random Feature subset**—which works similarly to Bagging, but instead of different subsets of samples, different subsets of features are used.
- **Noise addition**—the last method combines results of instance selection methods performed on different datasets with additional Gaussian noise.

For details of this ensemble methods see [4, 6].

2.2 Prototype Optimization Methods

The second group of data compression methods are prototype optimization algorithms. This group has two subgroups. The first one are prototype-based clustering methods, which represent clusters as a set of reference points (also called prototypes). The most famous example is a k-means clustering, but as it is the default RapidMiner algorithm, it is not reimplemented in the library. Instead the library contains:

- **Fuzzy C-means**—a modified version of k-means clustering, which incorporates fuzzy set theory to improve clustering results. As a result each data point belongs to every cluster with certain degree defined by the membership function. The final

crisp cluster association can be obtained by taking the cluster with the maximal membership function value.
- **Conditional Fuzzy C-means**—an extended version of Fuzzy C-means algorithm. It allows to incorporate external conditions to perform clustering. Its application to prototype optimization was discussed in [14].
- **Vector Quantization**—a very effective algorithm based on stochastic gradient descent, proposed by T. Kohonnen [11].

The second group is defined by the family of LVQ algorithms, which includes LVQ1, LVQ2, LVQ2.1, LVQ3, OLVQ. This group of methods is extended by the WLVQ algorithm with a modified cost function, which takes into account external conditions defined by a weight attribute. The weight can be obtained for example by the ENN algorithm or the Weight Transformation operator. The last implemented algorithm is SLVQ—a semi-supervised LVQ, which is a mixture of LVQ and VQ algorithms. If a training instance contains label information, then the LVQ scheme is used, but if the label information is missing, then the VQ training scheme is applied. The idea of using LVQ networks as a dataset compression method before training the SVM classifier was discussed in [5]. As an LVQ network has linear time and constant memory complexity it can be very efficiently used to reduce the computational complexity of the whole system.

3 Library Architecture

The library is written in Java and is available as a RapidMiner extension. The choice of RapidMiner as a data mining environment was based upon the following RapidMiner features:

- Very efficient internal data representation.
- High flexibility of building complex and advanced data mining processes using an intuitive graphical user interface.
- Popularity of RapidMiner in real world applications.
- Openness of the RapidMiner project (RapidMiner 5.x, which was the newest version when we begun working on the library is still available as open source and fully compatible with new version 6.x and 7.x. RapidMiner since version 6.x and 7.x became a commercial software with available free edition for education and research).

The internal data structure used to store datasets is based on ExampleSet class which is a *view* on so called ExampleTable—a structure, which directly stores the data. The concept of *views* was borrowed form data bases community and it allows to significantly reduce memory usage, as it does not require to duplicate the dataset before and after instance selection. Instead we have a view just on selected subset of examples (records or vectors) or features. This is crucial for processing massive datasets. To simplify implementation of new instance selection algorithms the

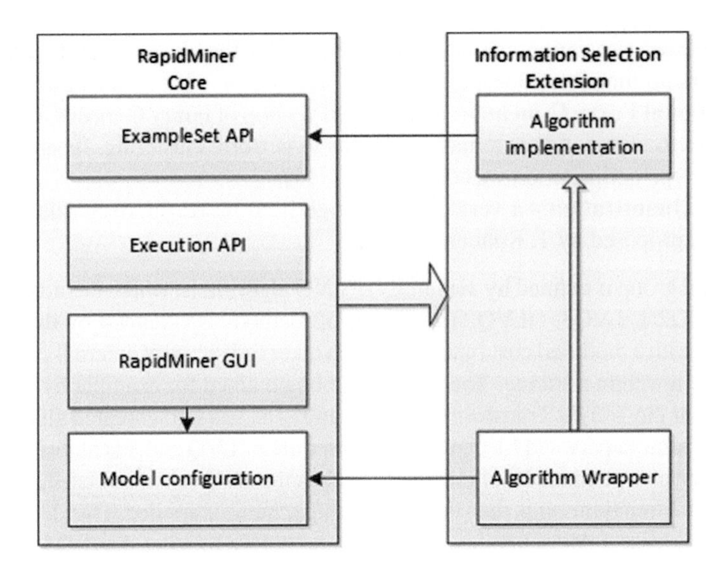

Fig. 2 Architecture of the Instance Selection Library

Information Selection and Data Compression library provides dedicated ExampleSet classes called SelectedExampleSet and EditedExampleSet, which for simplicity allow for binary switching on/off selected examples—similarly to the Matlab architecture. These two classes differ in the effect of instance switching. In the case of SelectedExampleSet the binary index takes effect after submission, while in the case of EditedExampleSet the binary switching has an immediate effect (Fig. 2)

The library is located in *ispr* package containing all the elements of the library. The *ispr* package includes:

- *example*—package dedicated to data structures.
- *selection*—package of instance selection operators—a set of wrappers of each of the algorithms:
  - *models*—package with pure algorithm implementations,
  - *meta*—package of ensemble of instance selection methods,
  - *generalized*—package that contains the generalized versions of ENN and CNN instance selection methods,
  - *weka*—an extension, which allows to call *Weka* instance selection methods.
- *optimization*—RapidMiner operators, which are wrappers of the algorithms defined in the packages:
  - *clustering*—with the clustering algorithms,
  - *supervised*—with the family of LVQ algorithms,
- *classifiers*—specific version of the *k*NNalgorithm,

- *loss*—various loss functions, which can be used for instance selection which are denoted as "support regression",
- *tools*—utility classes and some operators useful for assigning labels for non-supervised instance selection and optimization methods (such as labeling cluster prototypes),
- *container*—a set of data structures used to efficiently calculate nearest neighbors,
- *performance*—implementation of RapidMiner operators for performance measures such as compression level or clusters quality.

4 Usage Examples

The library was designed to utilize all the features provided by RapidMiner where one of the advantages is the creation of complex data mining processes in a graphical user interface with drag and drop. An example of the *Information Selection and Data Compression* library usage is presented in Fig. 3. Figure 3a shows scenario where after loading the data, the instance selection process is performed using the Generalized ENN which wraps the Decision tree. Thus the dataset is pruned and optimized to achieve the maximum performance of the final decision tree. The next example (Fig. 3b) shows another process, where after outliers removal done by the ENN algorithm, the LVQ network is used to find optimal location of prototypes for SVM training, where the LVQ network is initialized by the instances returned by the CNN algorithm.

Figure 4a, b shows examples of ensemble instance selection. The first one shows Bagging of CNN instance selection and the second one shows voting, where three different algorithms (ENN, CNN, RNG) are used to vote.

In general, an overview of the properties of selected instance selection and prototype optimization methods implemented in our library is presented in Fig. 5. The figure shows relations between compression and accuracy of an SVM classifier trained on the compressed dataset. The compression is defined as $\frac{n'}{n}$. It measures % of the examples from the training dataset, which remain after selection. Thus compression equal 1 means that all instances are selected (no compression at all). The baseline, which appears on the plot represents results obtained by the SVM classifier trained on the whole dataset without any compression. All results were obtained with the 5-fold cross-validation on the Spambase dataset obtained from the UCI repository. An SVM with Gaussian kernel was used and the parameters C and γ were optimized in cross-validation.

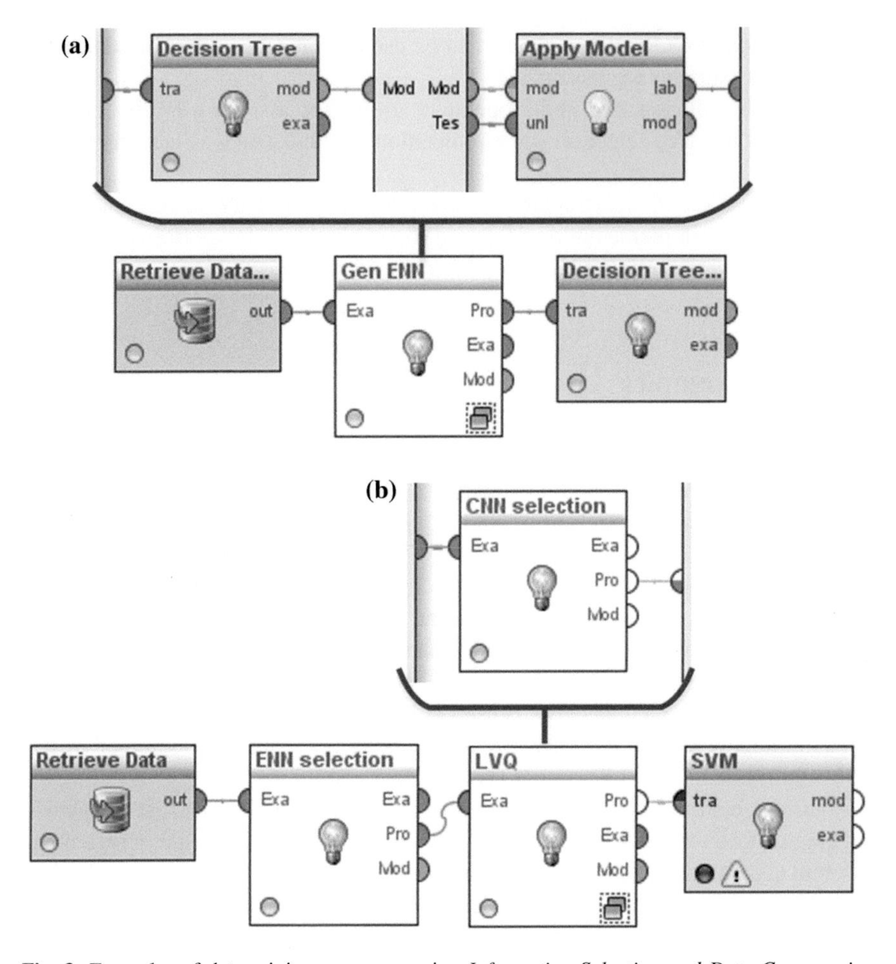

Fig. 3 Examples of data mining processes using *Information Selection and Data Compression* library. **a** Sample process with generalized ENN. Decision tree is used for inside instance selection. **b** Sample process with ENN and LVQ as prototype selection for SVM

As it can be observed, some of the algorithms provide small compression around 80 %. This group contains GE, ENN, RENN and All kNN but it has to be noted that these algorithms were designed for outlier elimination. Several instance selection algorithms like LVQ, RNG, CNN, RHCM and MC allowed for about 4-fold reduction of the dataset size, while reducing the accuracy only by 0.4 % in the case of LVQ and 0.5 % in the case of RNG. Such small decrease in accuracy is below the standard deviation.

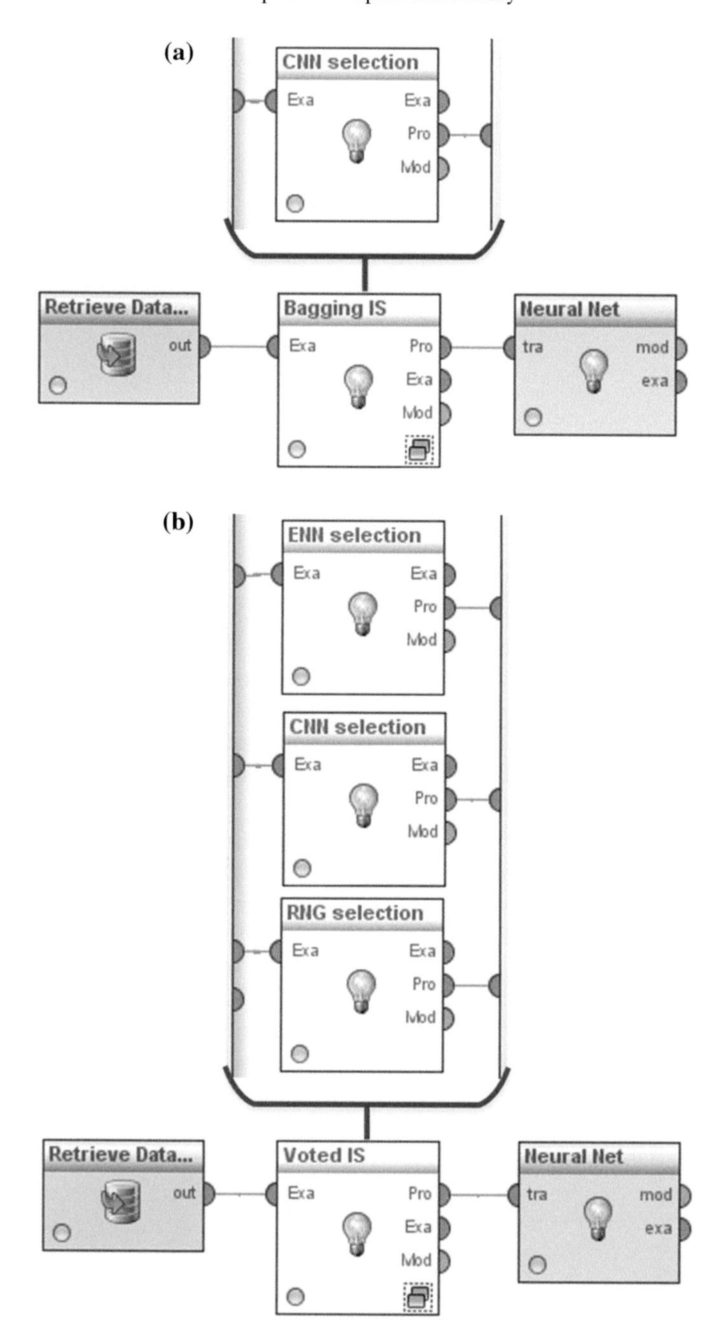

Fig. 4 Examples of data mining processes with ensembles of instance selection methods using *Information Selection and Data Compression* library. **a** Sample process with bagging for istance selection. **b** Sample process with voting used for instance selection

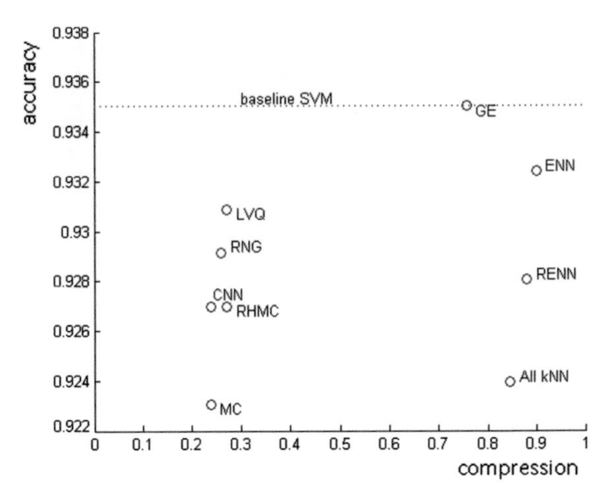

Fig. 5 Compression and accuracy obtained with various instance selection algorithms obtained on the Spambase dataset in 5-fold cross-validation, with SVM as the prediction algorithm

5 Conclusions

We have described the features and algorithms implemented in the *Information Selection and Data Compression* library. The library contains over 20 different algorithms, which can be used to filter and compress the dataset to provide better scalability of various data mining algorithms. As most of the original instance selection methods provide support only for classification tasks our library extends their functionality also to regression problems.

The library was designed as a plug-ins to the RapidMiner software, which at the beginning of the development was a fully open source project. Currently the core of RapidMiner stays open source and there is a free version for educational and research purpose. However, the more advanced versions of RapidMiner are available on a commercial license. RapidMiner versions up to 5.3 is still available on AGPL license. Our library is fully compatible with version (5.3, 6.x and 7.x).

The library is constantly being developed and new features are added. In the nearest future we plan to implement nonparametric noise estimation methods and additional instance selection algorithms. The standard version of the library can be obtained from the RapidMiner Marketplace and the most recent version can be accessed from the project website [3].

References

1. Asuncion, A., Newman, D.: UCI machine learning repository. http://www.ics.uci.edu/~mlearn/MLRepository.html (2007)
2. Aha, D., Kibler, D., Albert, M.: Instance-based learning algorithms. Mach. Learn. **6**, 37–66 (1991)
3. Blachnik, M.: Information selection and prototype-based rules. http://www.prules.org

4. Blachnik, M.: Ensembles of instance selection methods based on feature subset. IEEE Procedia Comput. Sci. **35**, 388–396 (2014)
5. Blachnik, M.: Reducing time complexity of svm model by lvq data compression. Submitted to ICAISC (2015)
6. Blachnik, M., Kordos, M.: Bagging of instance selection algorithms. LNAI **8468**, 40–51 (2014)
7. Cameron-Jones, R.: Instance selection by encoding length heuristic with random mutation hill climbing. In: Proceedings of the Eighth Australian Joint Conference on Artificial Intelligence, pp. 99–106 (1995)
8. Grochowski, M., Jankowski, N.: Comparison of instance selection algorithms ii. Results and Comments. LNCS **3070**, 580–585 (2004)
9. Garcia, S., Derrac, J., Cano, J., Herrera, F.: Prototype selection for nearest neighbor classification: taxonomy and empirical study. IEEE Trans. Pattern Anal. Mach. Intell. **34**(3), 417–435. http://dx.doi.org/10.1109/TPAMI.2011.142 (2012)
10. Hart, P.: The condensed nearest neighbor rule. IEEE Trans. Inf. Theory **16**, 515–516 (1968)
11. Kohonen, T.: Learning vector quantization. Self-organizing Maps, pp. 203–217. Springer, New York (1997)
12. Kordos, M., Rusiecki, A.: Improving mlp neural network performance by noise reduction. Lect. Notes Comput. Sci. **8273**, 140–151 (2013)
13. Kordos, M., Blachnik, M., Bialka, S.: Instance selection in logical rule extraction for regression problems. LNAI **7895**, 167–175 (2013)
14. Pedrycz, W.: Conditional fuzzy c-means. Pattern Recognit. Lett. **17**, 625–632 (1996)
15. Skalak, D.: Prototype and feature selection by sampling and random mutation hill climbing algorithms. In: Proceedings of the 11th International Conference on Machine Learning, pp. 293–301. Citeseer (1994)
16. Sánchez, J., Pla, F., Ferri, F.: Prototype selection for the nearest neighbour rule through proximity graphs. Pattern Recognit. Lett. **18**(6), 507–513 (1997)
17. Tomek, I.: An experiment with the edited nearest-neighbor rule. IEEE Trans. Syst., Man, Cybern. **6**, 448–452 (1976)
18. Wilson, D.: Assymptotic properties of nearest neighbour rules using edited data. IEEE Trans. Syst., Man, Cybern. **SMC-2**, 408–421 (1972)

Automatic Clustering Methods of Offers in an E-Commerce Marketplace

Anna Wroblewska, Bartlomiej Twardowski, Pawel Zawistowski and Dominik Ryżko

Abstract This work describes fully automatic clustering methods of offers in an e-commerce marketplace. Three different grouping approaches are proposed. We also designed and applied quality measures of clustering based on user-generated events. We assessed the proposed methods of clustering and compared them.

Keywords Clustering · E-commerce · Cluster quality

1 Introduction

In a modern e-commerce marketplace, precise description of offers is not always possible. Data provided by external vendors and sellers is not consistent, even if the same physical products are concerned. Particular descriptions are often incomplete, parameters are filled with default or incorrect values. Performing any kind of analysis or running algorithms (e.g. recommendations) on such data may lead to poor results. One solution to this problem is to map the offers onto a more "stable" representation which can become the basis for further analysis.

Recommender systems, which provide offer suggestions to users, are one of the typical services, which rely on offer definitions. More personalized offer suggestions make for better recommendations. The quality of recommendations is measured using business performance indicators like GMV or CTR values. Gross merchandise volume (GMV) indicates a total sales value for merchandise sold through a particular marketplace over a certain time frame. Click-through rate (CTR) is a way of measuring success of e.g. recommendations or email ad campaigns by calculating the ratio of the number of clicks on the generated links to the total number of recommendations or ads viewed by the users.

A. Wroblewska · B. Twardowski · P. Zawistowski · D. Ryżko (✉)
Allegro Marketplace Intelligence, Allegro Group, Warsaw, Poland
email: d.ryzko@ii.pw.edu.pl

A. Wroblewska · B. Twardowski · P. Zawistowski · D. Ryżko
Warsaw University of Technology, Warszawa, Poland
e-mail: Anna.Wroblewska@allegro.pl

D. Ryżko et al. (eds.), *Machine Intelligence and Big Data in Industry*,
Studies in Big Data 19, DOI 10.1007/978-3-319-30315-4_13

Allegro.pl is the largest marketplace in Eastern Europe. It provides a platform for individuals and business to buy and sell on-line. In Poland it has over 7 million active users and over 38 million offers. Clustering of offers in Allegro service is a very important task because of the lack of product definitions in the service. We can only group offers into more meaningful clusters or create any kind of mapping of offers to a product catalogue which can be treated also as clustering. The definition of clustering meaningfulness is very vague. Looking for offers in the service one can see a listing of offers regarding often the same products but sold by different vendors. They often are described a bit differently in terms of title, attributes and an offer photo. Thus we have some constraints and we can only assess the meaningfulness of offers (regarding associated products and their features) in the terms of data and tools available in our platform.
Hypothesis The goal of our research was to design and implement an effective method for offer clustering based on explicit assignment rules. The result of clustering service can be used by other marketplace services (e.g. ads, recommendations) as well as to do ad-hoc product based analysis. In order to measure the performance of our new clustering methods we proposed a new quality measure based on user-generated events. This measures are also used to benchmark newly created offer clustering algorithms against existing ones. In our work, we benchmarked our two rule based, language independent methods an against existing legacy offer tagging system.

In the following sections we make a review of offer grouping methods and describe our methods of offer clustering (Sect. 2) and state the same about the quality measures (Sect. 3). We show the results, discuss them (Sect. 4) and draw conclusions in the last part of the article.

2 Offer Clustering

The problem of items grouping is a classics in the field of computer science. There are a lot of methods for clustering, especially in a 2-dimensional space. However, our space is not that simple. Offers have different parameters in many dimensions and the space of particular parameters are very difficult to compare between each other.

Offers have titles, attributes, prices (or start prices when they are auctions), delivery information and longer descriptions which can be texts as well as pictures. All of this information are provided by sellers, but not all are required. This results in many unfilled information especially about offer attributes. Additionally, we have a category tree up to 8 levels deep. Offers have to be assigned to it by sellers.

A solution that copes with such a difficult heterogeneous data is described in [2]. This method is used by eBay that has a similar domain of offers as the Allegro marketplace. In their solution an item/offer is the basic unit of transaction, embodied by a set of unstructured and heterogeneous data. This includes auction title, description, attribute name-value pairs, price. The item data can be then categorized into three types: (1) binary variables for term occurrences in textual data (the best examples are offer title and description, i.e., $b = (b_1, b_2, \ldots, b_V)$ where V is the size of vocabulary); (2) categorical variables for indexed attribute values such as brand and color, i.e., $c = (c_1, c_2, \ldots, c_U)$ where U is the number of attributes; and (3) continuous variables for numerical data such as price, i.e., $g = (g_1, g_2, \ldots, g_S)$ where S is the dimension of numerical feature space. A so called latent product (group of offers) is represented as a vector of Bernoulli success probabilities, multinomial parameters, and Gaussian means: $z = (p_1, p_2, \ldots, p_V, \theta_1, \theta_2, \ldots, \theta_U, \mu_1, \mu_2, \ldots, \mu_S)$. The way of generating the latent products is to use EM (Expectation Maximization) algorithm. The belongingness of an item to a latent product is stochastically modeled as the following process:

- For binary variables: $b_v \sim Binom(p_v), \forall v \in 1, \ldots, V$;
- For categorical variables: $c_u \sim Mult(\theta_u), \forall u \in 1, \ldots, U$;
- For continuous variables: $g_s \sim \mathcal{N}(\mu_s, \sigma^2), \forall s \in 1, \ldots, S$.

Given a latent product z_k, the likelihood of an item x_i belongs to this product is:

$$p(x_i \mid z_k) = \prod_{v:b_{iv}=1} p_{kv} \prod_{v:b_{iv}=0} (1 - p_{kv}) \prod_{u} \theta_{ku} \times \prod_{s} \frac{1}{\sqrt{2\pi\sigma^2}} \exp\left(\frac{(g_{is} - \mu_{ks})^2}{2\sigma^2}\right), \quad (1)$$

where Θ_{ku} is the multinomial probability corresponding to the outcome c_{iu}.

Nevertheless, this method generates latent products that are item clusters that cannot be defined in an explicit and concise way, e.g. as rules having predicates based on the item data. In this method we need item to latent-product mapping.

Another method of grouping entities is MinHashing [3] algorithm. It is a probabilistic clustering method that assigns a pair of users to the same cluster with probability proportional to the overlap between the set of items that these users clicks. MinHashing is based on grouping similar entities. The similarity is defined as Jaccard coefficient. In example for two users: u_i, u_j is defined as the overlap between their item sets given by the formula:

$$S(u_i, u_j) = \frac{|C_{u_i} \cap C_{u_j}|}{|C_{u_i} \cup C_{u_j}|}. \quad (2)$$

The Jaccard coefficient takes values between 0 and 1. The corresponding distance function $D(u_i, u_j) = 1 - S(u_i, u_j)$ is a metric.

MinHashing can be used to similar domain as in our problem. We can measure the similarity of offers based on overlapping sets of attributes, words from offer name and discrete price values. However, similarly as in the previous method, clusters generated this way have no explicit definitions and are harmed by missing values what is in our case the common issue.

Another similar area to our research is document clustering, where features of documents are generated from frequent stemmed terms that occur in analyzed document corpus [1, 8, 15]. Vector space model that describes documents are binary variable that indicate presence or absence of any term in the feature vector [11] or the space definition is based on tf-idf measure [15]. The measure of similarity between documents is often based on Jaccard measure (also Tanimoto similarity) or cosine similarity where directions of feature vectors are compared.

Other methods of document clustering use concepts associated with documents and concept representation of documents (bag of concepts) [13]. The concept representation is for example result of wikification of documents [6, 7].

Summing up, a data which supply the clustering procedure in our research of offer clustering can be derived from:

- only offer content, such as: titles, attributes and their values, linkage to other external data sources, such as Wikipedia can also be used when available,
- user-generated data, such as co-views of offers, i.e. offers views in one session of a user or search phrases and their linkage to particular offers,
- hybrid methods, mixed methods based on content and user-generated data.

The main goal of our work is to generated offer clusters based on structural information. The best outcome is to generate explicit rules that assign new offers to the clusters. The methods is independent of the offer descriptions language. Then we measure the efficiency of our clustering methods based on user-generated data. Additionally we compare the results of our new clustering methods with the old grouping in our system based on offer tagging when the clusters has no explicit definitions based on structural information in our service.

In the following we name offer groups generated by our methods as metaitems, clusters or offer groups.

2.1 Ars Clustering

Ars-Clustering (Allegro Recommendation System Clustering) is an offer clustering used by our legacy recommendation system. It is based on tagging information so every offer is assigned a tag which matches that offer best. Due to implementation architecture restrictions, only a single tag could be selected for a given offer. Imposing such a constraint caused the tag's value schema to evolve from a simple text value to a collection of recognised offer parameters like: manufacturer, model, colour, size, year of production, etc. As a source of information for the tagging process offer's name and attributes are utilised (Fig. 1). If tag disambiguation is needed (e.g. different meaning of the word between different categories of offers) the clustering process uses the most frequent tags in selected categories based on user search and click events.

The main drawback of Ars-Clustering is a language dependency. It uses natural language processing tools in order to prepare text and get the base form of the words (stems). This narrows down its usage to only the Polish part of the marketplace. Newer clustering methods try to overcome this shortcoming.

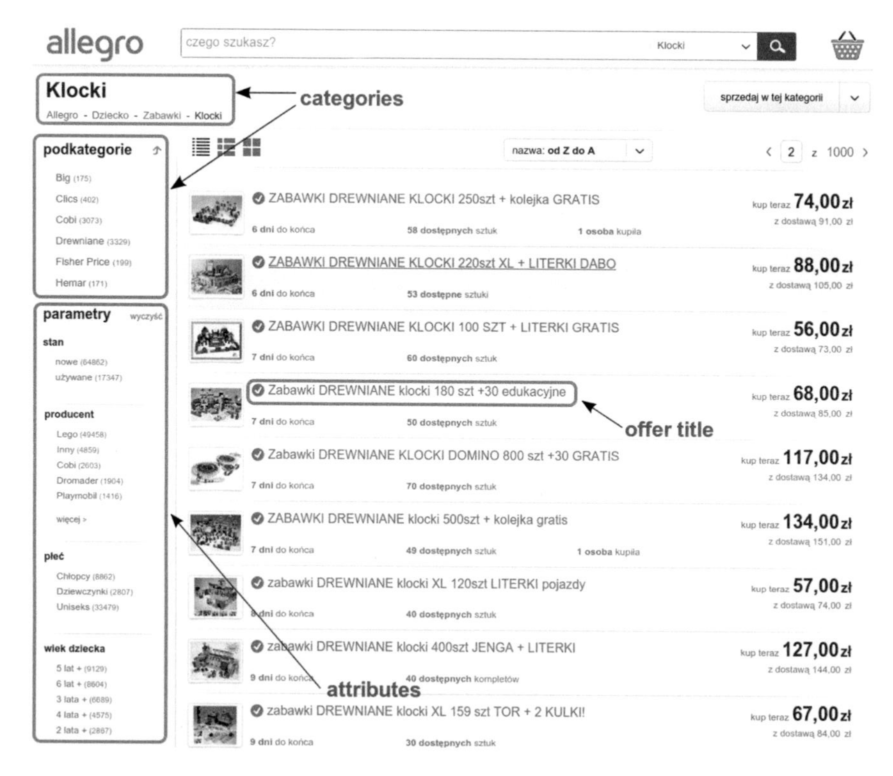

Fig. 1 Categories, attributes and offer titles in Allegro website

Despite the fact that Ars-Clustering mechanism is considered to be a part of our legacy system, it performs well. That is the main reason why we use it as our baseline benchmark to evaluate new approaches.

2.2 *CAP Clustering*

CAP (Category and Predicates) is a clustering method which groups Allegro offers using only the category tree and offers' attributes. In the first step we combine leaf categories containing too few active offers (the exact threshold value is a parameter of the method). During this process two small categories may be merged together only if they have the same parent category.

After processing category tree leafs, categories with the number of offers above the threshold are grouped together in an "active" list. Then, each element from this list is divided by the values of one of the attributes assigned to offers the list contains. This partitioning attribute is chosen on the basis of the number and sizes of subgroups it generates—ideally such an attribute divides the offer set into many equally sized subsets. The newly created subgroups are added to the end of the "active" list if they are large enough. Then the processing continues until all elements from the list have been processed.

When the procedure described above terminates, the ending situation is that each offer is assigned to a metaitem. The metaitem itself is defined by a category leaf and a sequence of [attribute, value] pairs. This results in a set of rules which are used for further offer to cluster classification.

2.3 *Mojito Clustering*

Mojito is the successor of CAP which split offers into clusters using the category tree, attributes and price shelves defined for offer subcategories (Figs. 2 and 3).

Our price shelves split each category price range into two or three shelves that contain cheap, moderetely priced and expensive offers. In order to create price shelves for a given category we use simple thresholds dependant on mean and standard

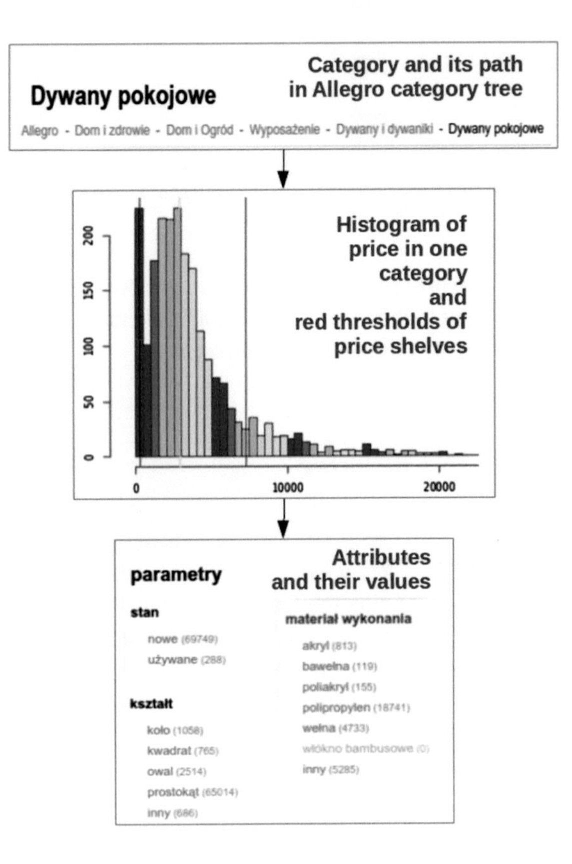

Fig. 2 Scheme of an algorithm and source data for generating mojito metaitems

Fig. 3 Example of a mojito metaitem definition

Category_path = **Dom i Ogród/Wyposażenie/ Dywany i dywaniki/Dywany pokojowe**

Price_shelf = **['{name:price, range: 2 over 3, min:200, max:1133.5}']**

Attribs = **['11525_Materiał wykonania=4_polipropylen', '15795_Kształt=2_prostokąt', '11323_Stan=1_nowy', '11508_Antypoślizgowość=2_nie']**

deviation of offer prices in that category. In the clustering algorithm, price shelves are treated as special attributes—a price shelf is always on the first place in the sequence of attributes defining a metaitem. Apart from that, the clustering procedure remains unchanged.

In comparison to CAP, Mojito also removes some attributes from the cluster generating procedure. The usefulness of attributes was measured on the basis of their frequency in CAP metaitem definitions—the least frequent ones have been omitted.

3 Quality Measures

Assessing the quality of the proposed offer grouping methods is not an easy task. One possibility is to treat the created offer groups as clusters and apply one of the available clustering validity techniques to perform evaluation. Such methods generally can be divided into three groups [5]: external, internal and relative criteria.

External validity criteria evaluate the clusters with respect to a pre-specified (external) structure imposed on the data set—e.g. class labels assigned to each data point. Among others, external criteria include normalised mutual information [14], entropy [18], purity [18] and F-measure [18].

Relative validity criteria work by comparing clustering results obtained by e.g. using the same algorithm with different parameter settings, thus given a set of distinct data partitions such criteria give the possibility to rank them. Similarly to the external and internal cases, there are many relative validity criteria available in the literature—[17] provides a list of 40 such measures.

To evaluate the grouping methods proposed in this paper we decided to use an external-type approach to confront our clusters with naturally occurring groupings. A straightforward procedure would be to label the offers with actual product names and verify whether each cluster represents a single product. Unfortunately this was not feasible in our case due to lack of product definition in our data. As a consequence we incorporated actual website traffic in the cluster evaluation process and designed two validity criteria based on sequences of user interactions with Allegro offers. These measures, which are introduced in two following sections, allowed us to empirically verify whether the items within each of the clusters are actually perceived as similar by Allegro users.

3.1 Cluster Validity Measure Based on User View Sequences

The first proposed metaitem quality measure is based on analysing the stream of user offer views divided into distinct 15-min slices. The underlying assumption is that, on average, during a time slice a given user is looking for a specific product so the offers viewed in such a slice should be assigned to the same metaitem. This means that metaitems are considered "better" when they contain offers which are frequently viewed together. Consequently "worse" metaitems consist of offers viewed separately.

Let q_0 denote the described measure. To calculate it, first a sequence of triplets [timestamp, user, viewed offer] needs to be obtained from website traffic. This sequence is then divided into 15 min slices for each distinct user and individual offers are mapped onto the corresponding metaitems. q_0 is then formally defined as follows:

$$q_0(M) = \frac{1}{SU} \sum_u \sum_s \frac{|\alpha_{s,u} \cap M|}{|\alpha_{s,u}|} \tag{3}$$

where

- M—a set of offers corresponding to a single metaitem,
- $s = 1, \ldots, S$—a time slice,
- $u = 1, \ldots, U$—a user,
- $|\alpha_{s,u}|$—the number of occurrences of all metaitems in time slice s for user u,
- $|\alpha_{s,u} \cap M|$—the number of occurrences of offers from metaitem M in time slice s for user u.

3.2 Cluster Validity Measure Based on User Decision Diversity

The second approach to measure quality of an offer clustering is similar to the q_0 measure. The new measure (q_1) is based on user's decision diversity observed before offer purchase. Additionally to sequence of offer views, q_1 incorporates user transaction data. Our assumption is that sessions where offer acquisition occurred are focused on specific products—users are trying to find best fitted offers to buy. Thus, the diversity of offer page views in such sessions is low and most of them are connected with single purchase decisions. In our case, this fact is utilised to validate metaitem clustering quality.

Measure q_1 based on user decision diversity is defined as follows:

$$q_1(M) = \frac{1}{S_M U} \sum_u \sum_{s_M} \frac{|t_{s,u}|}{|\alpha_{s,u}|} \tag{4}$$

where

- M—a set of offers corresponding to a single metaitem,
- $s_M = 1, \ldots, S_M$—user sessions which end up with a transaction with at least one item belonging to metaitem M,
- $u = 1, \ldots, U$—a user,
- $|t_{s,u}|$—the number of distinct metaitems in transaction basket in session s for user u,
- $|\alpha_{s,u}|$—the number of distinct metaitems in session s for user u.

4 Results and Discussion

To perform the evaluation of the proposed offer grouping methods (ars-clustering, cap and mojito) we used data collected during a 3-month period at the turn of 2014 and 2015. Each offer present in that set was assigned to appropriate metaitems for all three methods. As in the previous section, a metaitem here is a set of offers that comply to the defining rules of that particular metaitem.

This set was then used to calculate values of both proposed validity measures.

4.1 *Results for the q_0 Measure*

An analysis of the numbers of metaitems appearing in the time slices reveals that about 0.7 % of metaitems appear only in a single slice during the entire 3-month period. Furthermore, about 80 % of these metaitems have only one user view in that slice. Thus they are very rare and we came to the conclusion that we would analyze these metaitems in order to catch the definitions and attributes that are probably incorrectly set.

Histogram of metaitems sizes is show in Fig. 4. Ars-clustering metaitems are a bit smaller on average, and mojito metaitems are more diversified in size and have more smaller metaitems than cap metaitems. The reason for this is that ars-clustering divide offers with tags derived from offer titles and provide the smallest granularity among all metaitem kinds. Mojito metaitems have all features of cap metaitems and they divide offers into price shelves additionally.

Histogram of q_0 measure values for our three types of metaitems is shown in Fig. 5. The differences of distributions of q_0 measure between the 3 types of metaitems is statistically significant. To investigate this we used Kolmogorov-Smirnov test for equality of distributions and Wilcoxon rank sum test for equality of means at the 0.95 significance level.

From Figs. 4 and 5 it can be noticed that in one time slice user viewed offers that are characterised by similar tags (ars-clusters) and comparable prices (mojito features). Ars-clustering achieved the best results according to q_0 measure, so we can

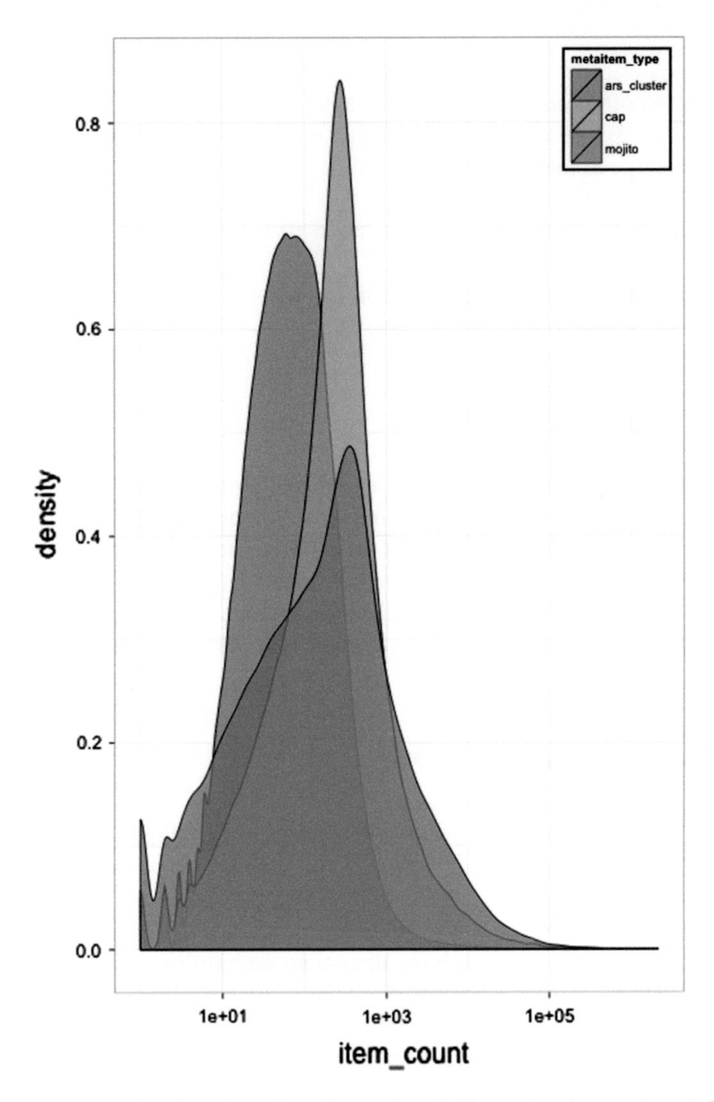

Fig. 4 Histogram density of metaitem sizes (the number of offers assigned to metaitem definitions)

conclude that they are the most coincident with user views. Mojito are a bit worse than ars-clustering, but are significantly better than cap metaitems. On the other hand, q_0 values aggregated over all kinds of metaitems have mean values below 0.5, which suggests that this measure is not clearly coincident with user views. Users do not navigate through the offers in our service considering all attributes, prices and tags utilised by ars-clustering.

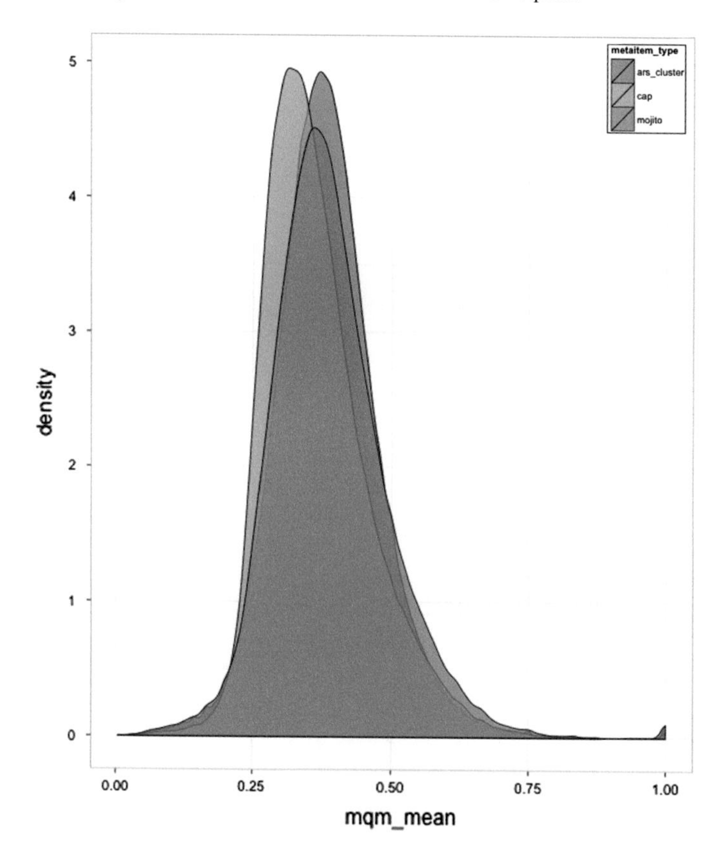

Fig. 5 Histogram density of q_0 measure for our three types of metaitems

4.2 Results for the q_1 Measure

Based on the same 3 month time period on which the q_0 measure was evaluated the calculation of q_1 was performed. Taken sample has $U = 5269407$ different users which bought at least one item on Allegro.pl. Number of sessions taken part into calculation differs for specific metaitem type. This is the result of different classification to cluster approach. The whole number of sessions S was: for Ars-clustering $S = 1468315128$, for CAP $S = 1272950713$ and for mojito $S = 1198105832$. As expected, the most granular Ars-clustering has the highest number sessions.

The density histogram of q_1 measure is presented in Fig. 6. The best result achieve on our evaluation set for q_1 measure have Ars-clustering, then mojito and CAP. The result between CAP and mojito is expected. Mojito clustering added better granularity in clustering based on price discretization to specified ranges. But still, Ars-clustering based on tagging approach and the finest granularity wins in q_1 comparison.

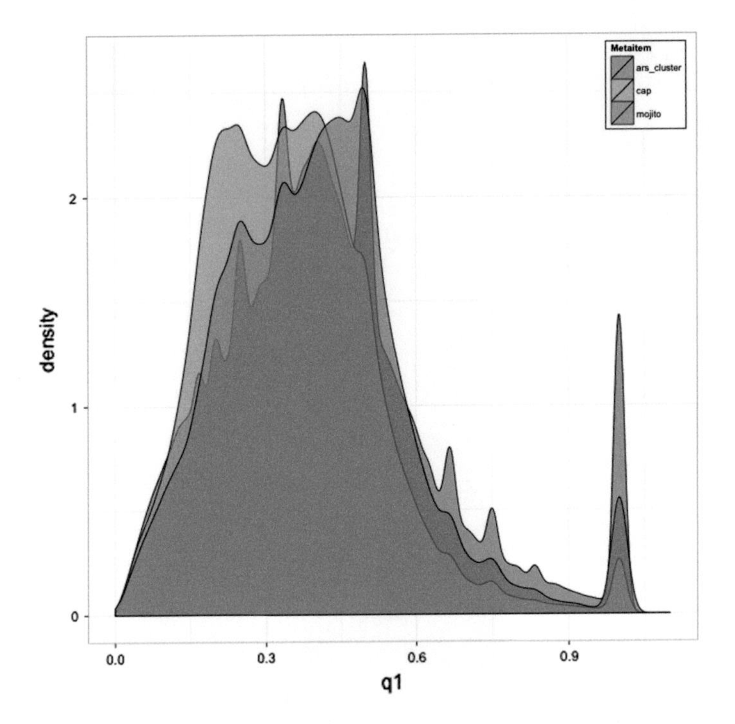

Fig. 6 Histogram density of q_1 measure for all three types of metaitem

What needs a word of explanation is the high peek around 1.0 value. This is the situation when the same number of metaitem's offers is bought as viewed. Most of this sessions are transaction which finalized the buying process of previously chosen items. In Allegro.pl website or mobile application it can be done by saving offers to favourites or by creating and saving cart.

5 Conclusions

We deployed both presented methods of clustering in Allegro service. This results in better understanding of offers created by users on our marketplace. More stable CAP and Mojito clusters can be treated as approximate product definition equivalents. This is crucial for some insight business analysis and useful for systems based on products like advertisements and recommendations.

Both of the presented methods result in clear cluster definitions based on rules. Such expression is helpful from analysis perspective. Moreover, it enables a straightforward and efficient implementation of offer assignment to an appropriate cluster. In order to do this, an appropriate rule needs to be found by traversing the tree structure defined by categories and attributes. In contrast to clustering methods based on distance measures this results in effective and fast assignment. This is important as in our environment clustering has to be done as soon as possible in online manner.

What is more, the definitions of clusters are stable. This is different from centroid or density methods in which cluster approximations may change due to new data acquisition.

On the other hand, the main issue with CAP and Mojito clustering is its inability to adopt automatically to categories' and attributes' changes. In order to properly classify offers in newly added category or using new attributes new definitions of clustering rules need to be provided. Otherwise, most of the offers with new attributes will be classified in generic clusters like: "Accessories", "Others", etc. Nevertheless it does not happen often, since changes in marketplace category tree are unusual.

While using the clustering service by other systems, new ideas for better product approximation and new requirements for the data emerge. However, providing new methods of clustering to production use without any prior evaluation is risky. To minimize a possibility of data corruption we started to measure how well clustering methods perform. What is more, having some quality measures we can compare methods and benchmark new clustering approach with the existing legacy system (Ars-Clustering). That results in efforts to introduce cluster validity measures q_0 and q_1.

Futures plans include taking into account evaluation how other services in our marketplace ecosystem benefit from the new clustering approaches. One of the main system depending on cluster quality is the recommendation platform. Thus our next investigation should take into consideration results of A/B testing. An experiment is scheduled to run for selected scenarios where different groups will receive product suggestions based on different clustering types.

In our future work we also consider using NLP techniques to get more insight information from free text offer descriptions. Although this is a language dependent approach, it is still necessary in some categories where no product definition exists and offers are unique. This is also proven by the notable result of our legacy system Ars-clustering which extracts data only from offer names and attributes. We would like to expand this approach and analyse the entire offer where a large number of information resides in free form descriptions made by users. Moreover, we are considering using attributes extracted from pictures as an input for our clustering methods. Many offers are created automatically by dedicated importer tools, which results in most of them having similar or almost the same description. What is different from the buyers perspective is that the pictures presenting product's vary, e.g. in colours or textures. Thus, extracting such visual features can provide additional information and better clustering results.

References

1. Agrawal, R., Phatak, M.: A novel algorithm for automatic document clustering. In: IEEE 3rd International Advance Computing Conference (IACC), pp. 877–882 (2013)

2. Chen, Y., Canny, J.F.: Recommending ephemeral items at web scale. SIGIR (2011)
3. Das, A., Datar, M., Garg, A.: Google news personalization: scalable online collaborative filtering. WWW (2007)
4. Hubert, L., Arabie, P.: Comparing partitions. J. Classif. **2**(1), 193–218 (1985)
5. Halkidi, M., Batistakis, Y., Vazirgiannis, M.: On clustering validation techniques. J. Intell. Inf. Syst. **17**, 107–145 (2001)
6. https://github.com/dbpedia-spotlight/dbpedia-spotlight/wiki/Introduction
7. Milne, D., Witten, I.H.: Learning to link with Wikipedia. In: Proceedings of the 17th ACM Conference on Information and Knowledge Management. CIKM '08, pp. 509–518. ACM, New York, USA (2008)
8. Nourashrafeddin, S.N., Milios, E., Arnold, D.V.: Interactive text document clustering using feature labeling. In: Proceedings of the 2013 ACM Symposium on Document Engineering. DocEng '13, pp. 61–70. ACM, New York, USA (2013)
9. Nourashrafeddin, S., Milios, E., Arnold, D.: An ensemble approach for text document clustering using Wikipedia concepts. ACM DocEng (2014)
10. Sharma, S.: Applied Multivariate Techniques. Wiley, New York (1996)
11. Shameen, M., Ferdous, R.: An efficient k-means algorithm integrated with Jaccard distance measure for document clustering. First Asian Himalayas International Conference on Internet, AH-ICI, pp. 1–6 (2009)
12. Schafer, J.B., Konstan, J., Riedl, J.: Recommender systems in e-commerce (1999)
13. Spanakis, G., Siolas, G., Stafylopatis, A.: Exploiting Wikipedia knowledge for conceptual hierarchical clustering of documents. Comput. J. **55**(3), 299–312 (2015)
14. Strehl, A., Strehl, E., Ghosh, J., Mooney, R.: Impact of Similarity Measures on Web-page Clustering. W. on Art. Intel. for Web Search (AAAI), pp. 58–64 (2000)
15. Shi, K., Li, L., He, J., Liu, H., Zhang, N., Song, W.: A linguistic feature based text clustering method. IEEE International Conference on Cloud Computing and Intelligence Systems (CCIS), pp. 108–112 (2011)
16. Twardowski, B.: Long tail problem and implicit user feedback in e-commerce marketplace recommender systems (2014)
17. Vendramin, L., Campello, R.J., Campello, G.B., Hruschka, E.R.: Relative clustering validity criteria: a comparative overview. Statistical Analysis and Data Mining, vol. 3, pp. 209–235. Wiley, New York (2010)
18. Xiong, H., Wu, J., Chen, J.: K-means clustering versus validation measures: a data-distribution perspective. IEEE Trans. Syst., Man Cybern. **39**, 318–331 (2009)

Application of Machine Learning Algorithms for Bitcoin Automated Trading

Kamil Żbikowski

Abstract The aim of this paper is to compare and analyze different approaches to the problem of automated trading on the Bitcoin market. We compare simple technical analysis method with more complex machine learning models. Experimental results showed that the performance of tested algorithms is promising and that Bitcoin market is still in its youth, and further market opportunities can be found. To the best of our knowledge, this is the first work that tries to investigate applying machine learning methods for the purpose of creating trading strategies on the Bitcoin market.

1 Introduction

Bitcoin is a peer-to-peer, decentralized electronic cash protocol. It is based on the proof-of-work algorithm that uses cryptographic hash function for confirmation of subsequent blocks of transactions. Block's approval is realised by miners. Skipping the details, the general idea behind is to find such a 32-bit field (called *nonce*) that added to the block will result in a value of a hash function with n leading zeros. Miners compete between each other to earn the incentive for finding such a value. In order to achieve this goal, they have to employ the computational power bearing the costs of hardware and electricity. The more miners are engaged in the described process, the harder it is to find the required value of hash function as more leading zeros are required in order to accept the block. The measure describing this is called *difficulty*, and it is self-regulated by the protocol.

Bitcoin was proposed by Satoshi Nakamoto in 2008 [1]. Since then the whole Bitocoin ecosystem has been growing rapidly. Since then it also raises questions about some fundamental issues it promises. One of them is the anxiety about the privacy and anonymity. More on this can be found in [2, 3]. For a long time, Bitcoin's proof-of-work has been perceived as a safe to potential attacks. Recently, there have been some concerns about that as well [4].

K. Żbikowski (✉)
Faculty of Electronics and Information Technology, Institute of Computer Science,
Warsaw University of Technology, ul. Nowowiejska 15/19, 00-665 Warsaw, Poland
e-mail: kamil.zbikowski@ii.pw.edu.pl

D. Ryżko et al. (eds.), *Machine Intelligence and Big Data in Industry*,
Studies in Big Data 19, DOI 10.1007/978-3-319-30315-4_14

Despite some potential threats, Bitcoin community is growing and attracts more and more attention both from the academia and business. In recent years, plenty of Bitcoin exchanges have been launched. The main advantage of them is a simplification of trading and valuation of Bitcoin. As there is a market there also exist investors, speculators and hedgers ready to exchange bitcoins and fiat currencies. As it is done on traditional financial markets, one may be interested in applying automated trading strategies that would help gain reasonable return on invested funds. In this paper, we would like to test and compare several approaches from traditional markets applied to the task of automated trading on Bitcoin.

The remainder of this paper is organised as follows. Section 2 briefly describes the concept of Support Vector Machines and Volume Weighted Support Vector Machines. We also provide information about indicators used for the purpose of presented experiments. In Sect. 3 we provide information about data that was used for computations. In Sect. 4 we analyze experiments' results. Finally, in Sect. 5 we make some concluding remarks.

2 Prediction Models

In order to investigate the existence of profitable automated trading algorithms, we propose to use some of the most well-known approaches. First, we would like to verify basic exponential moving average (EMA) with some simple enhancements. Then, we would like to apply a much more sophisticated method that directly employs machine learning methods like support vector machines (SVM) in combination with box theory. Last but not least, we would like to determine whether some improvements in the machine learning algorithm, validated against data from other financial markets, will also perform well.

2.1 Exponential Moving Average

First model that we use is exponential moving average (EMA) that is widely known technical analysis indicator. It assigns to the more recent quotations, higher weights basing on the assumption that they are more relevant than the older ones. EMA for the moment t is defined recurrently as follows:

$$EMA_t^n = (P_t - EMA_{t-1})\alpha + EMA_{t-1} \tag{1}$$

where P_t is the price for the moment t, α is defined as $2/(n+1)$ and the value of EMA_0 is set to P_0.

Signals to buy or sell are triggered when the $EMA_t < P_t$ or $EMA_t > P_t$ respectively. Some traders believe that there exist predefined values of n that suit for every case. We took a rather different approach. We applied walk-forward procedure [5] in order to re-optimise the model every 100 quotations. During this step, we conduct an extensive search using grid search procedure looking for the value of n for the next moving window basing on last 200 quotations.

2.2 Support Vector Machines with Box Theory

Support Vector Machine (SVM) is a machine learning algorithm that employs the risk minimization principle that was described in [6]. Its general idea is based on maximizing the margin of separating hyperplanes. This operation prevents from building models that are overfitted and directly lowers the error obtained on the test sample. The general penalty function is defined as follows:

$$\min_{\mathbf{w}\in\mathcal{H},b\in\mathcal{R}} \frac{1}{2}||\mathbf{w}||^2 + C\sum_{i=1}^{m}\xi_i, \tag{2}$$

subject to:

$$y_i(\mathbf{w}\mathbf{x_i} - b) \geq 1 - \xi_i, \tag{3}$$

$$\xi_i \geq 0, \tag{4}$$

where C is a penalty term, ξ_i are slack variables, x_i is the ith input value and y_i is the corresponding target value.

The model based on box theory and SVM was proposed in [7]. The box theory assumes that in order to obtain valid buy or sell signals the price should break through the oscillation box constructed over the specified period of time. The input of the model consists of several technical indicators and statistical measures gathered in Table 1.

2.3 Volume Weighted Support Vector Machines

The idea of Volume Weighted SVM was introduced in [5] and further investigated in [8]. It is based on the simple, yet powerful, assumption that latest quotations should have a greater impact while building a prediction model. To achieve this goal, there was proposed a slight modification in a basic SVM formula 2 that takes following form:

$$\min_{\mathbf{w}\in\mathcal{H},b\in\mathcal{R}} \frac{1}{2}||\mathbf{w}||^2 + \sum_{i=1}^{m}C_i\xi_i, \tag{5}$$

In 5, each C_i term determines the weight assigned to particular input x_i and takes the form:

$$C_i = v_i C, \tag{6}$$

where

$$v_t = \frac{\sum_{k=0}^{d} V_{t-k}}{\sum_{i=0}^{m} W_{t-i}}, \tag{7}$$

where V_{t-k} is the real transactional volume for the moment t, k depicts the delay period and d is the data length required for feature calculation.

In [8], the formula 5 was used in combination with several technical indicators. The list of indicators used as an input for VW-SVM for the purpose of the experiment presented in this paper is enlisted in Table 1.

3 Research Data

For the purpose of presented experiment, we used data gather from one of the largest Bitcoin exchanges in terms of average daily volume traded—Bitstamp (https://www.bitstamp.com). It is available via their API. We have merged all the quotations to the form of 15-min aggregates and as a price P_t we have applied the closing price of an aggregate from the moment t. The spectrum of selected data consists of 2221 aggregates. It starts on January 9th 2015 and ends on February 2nd 2015.

Transactions' costs vary depending on exchange and traded volume over time. We assumed transaction fee at the level of 0.2 % which seemed to be a reasonable choice.

4 Empirical Study

We applied a walk forward testing [5] procedure that is a recommended method for constructing machine learning models dedicated to time-series data. Each algorithm was able to open both short and long positions. That is not directly possible at the moment having an account open on the Bitstamp exchange. However, several competitors provide such a service for their customers. Therefore, we got to the conclusion that this decision was justified.

We run each of the models described in Sect. 2 separately using data that was collected according to Sect. 3. The summary of these experiments is collected in Table 2. We assume that initially, each of the algorithms holds a portfolio of 100,000 $.

In order to measure the degree of risk we provide a drawdown measure that is computed as a maximum loss from the highest value of the equity that algorithm was holding (expressed in percent). As an assessment of model's performance, we provide rate of return computed over the whole testing period. All results are compared to the

Table 1 Indicators used as the inputs of Box SVM and VW-SVM models (n denotes the data length used for computations; t is the index of the element in the time series)

Indicator	Formula	Box SVM	VW-SVM
Close price	P_t	+	+
Volume	V_t	+	+
Average price	$\mu_t = \frac{\sum_{k=0}^{t} P_{t-k}}{t}$	+	+
Standard deviation	$\sigma_t = \sqrt{\frac{\sum_{k=0}^{t}(P_{t-k}-\mu_t)^2}{t}}$	+	+
Relative strength index	$RSI_t^n = 100 - \frac{100}{1+RS_t^n}$ $RS_t^n = \frac{\sum_{k=0}^{n} U_k}{\sum_{k=0}^{n} D_k}$ $U_t = \begin{cases} P_t - P_{t-1} & \text{if } P_t \geq P_{t-1} \\ 0 & \text{otherwise} \end{cases}$ $D_t = \begin{cases} P_{t-1} - P_t & \text{if } P_t < P_{t-1} \\ 0 & \text{otherwise} \end{cases}$	+	+
Daily rate of return	$R_t = \frac{P_t}{P_{t-1}}$	−	+
%R Oscillator	$\%R_t^n = 100\frac{MaxP_t^n - P_t}{MaxP_t^n - MinP_t^n}$ $MaxP_t^n = \max\{P_t, P_{t-1}, \ldots, P_{t-n}\}$ $MinP_t^n = \min\{P_t, P_{t-1}, \ldots, P_{t-n}\}$	−	+
Average true range	$ATR_t^n = \frac{1}{n}\sum_{i=0}^{n} TR_{t-i}$ $TR_t = \max\{P_t^H - P_t^L, P_t^H - P_{t-1}^C, P_t^L - P_{t-1}^C\}$	−	+
Vortex	$V_t^{n(+)} = \frac{\sum_{k=0}^{d}(P_{t-k}^H - P_{t-k}^L)}{\sum_{k=0}^{d} TR_{t-k}}$ $V_t^{n(-)} = \frac{\sum_{k=0}^{d}(P_{t-k}^L - P_{t-k}^H)}{\sum_{k=0}^{d} TR_{t-k}}$	−	+
On balance volume	$OBV_t = \begin{cases} OBV_{t-1} + V_t & \text{if } P_t > P_{t-1} \\ OBV_{t-1} - V_t & \text{if } P_t < P_{t-1} \\ OBV_{t-1} & \text{if } P_t = P_{t-1} \end{cases}$	−	+

Table 2 Backtesting results for EMA, Box SVM and VW-SVM models

	R (%)	DD (%)
EMA	−59.11	64.30
Box SVM	10.58	25.94
VW-SVM	33.52	22.88
Buy and Hold	4.86 (28.10)	30.90

R is the rate of return over the whole testing period, DD is the maximum drawdown experienced by the algorithm

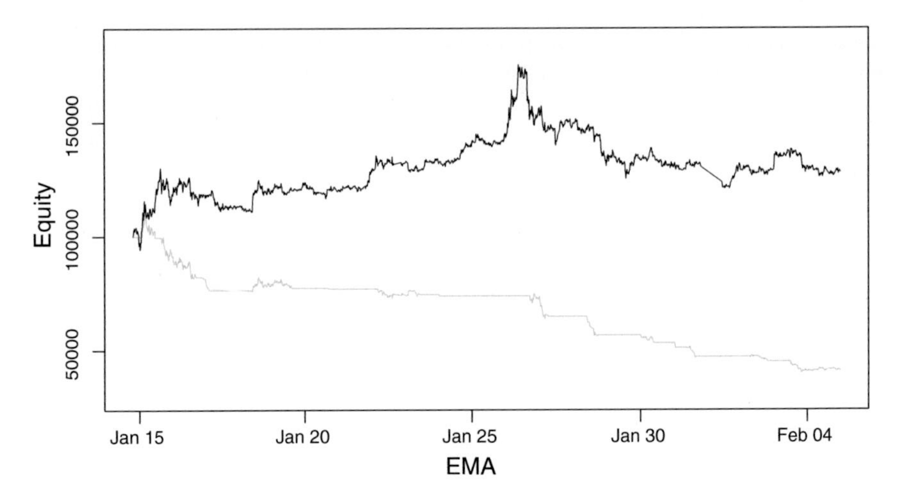

Fig. 1 EMA results. *Black*—buy & hold strategy, *gray*—EMA performance

so-called buy and hold strategy that assumes opening a long position at the beginning of available data and holding up to the end of data. Each time the position is open by the algorithm it engages the whole capital that it is holding at the particular moment. It is rather a weak position size management. However, it is justified if we compare trading result to buy and hold reference strategy.

Simple exponential moving average model lost 59.11 %. As it can be seen in the Fig. 1 the equity line is declining throughout the whole tested period of time. The drawdown of this model reached 64.30 %. This strategy obtained the worst result in terms of both maximum drawdown and the rate of return. The buy and hold strategy performed significantly better. From this outcome we may conclude that even for such an immature market like it is in case of Bitcoin, simple trading strategies based on technical analysis are not profitable.

First of the advanced strategies that we were testing, namely Box SVM, achieved much better results compared to EMA. Its rate of return was 10.58 % which is 5.72 % improvement over buy and hold (see Fig. 2). In terms of the maximum drawdown, it did not cross the level of 25.94 % which still denotes a high risk of investment. However, compared to buy and hold strategy, it achieved 4.96 % improvement.

The second strategy that is based on SVM classifier, namely Volume-Weighted SVM, managed to achieve the rate of return of 33.58 %. Its equity curve is shown in the Fig. 3. It is the highest return ratio among analyzed models. It is 28.66 % improvement over buy and hold strategy and 23 % over Box SVM. Moreover, VW-SVM exposure to the risk measured by the maximum drawdown was the lowest. It was better than buy and hold by 8.02 % and by 3.14 % than Box SVM model.

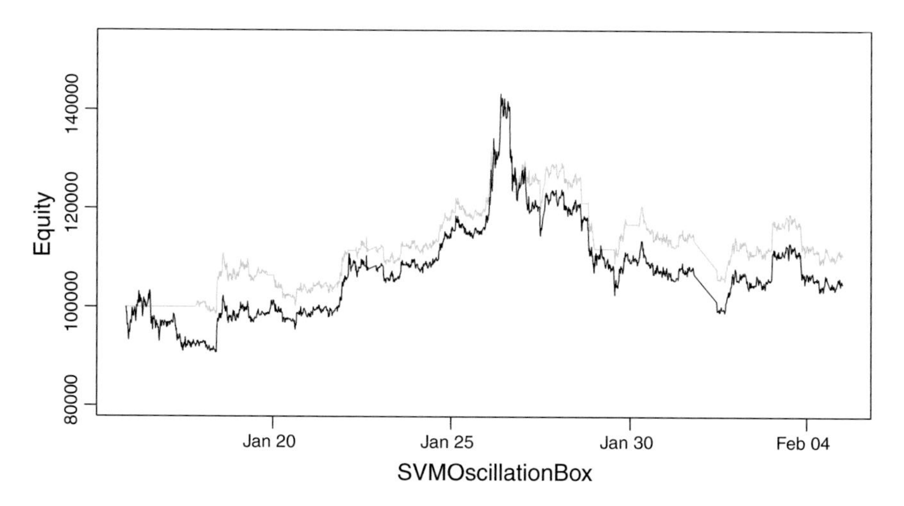

Fig. 2 Box SVM results. *Black*—buy & hold strategy, *gray*—box SVM performance

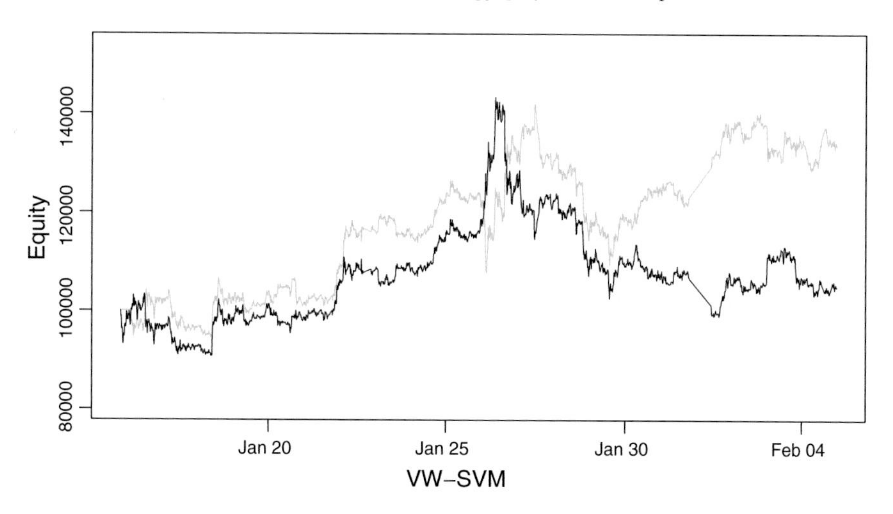

Fig. 3 VW-SVM results. *Black*—buy & hold strategy, *gray*—VW-SVM performance

5 Conclusion

In this paper, we compare three different algorithms that are applied to the task of trading on Bitcoin exchange. One of them was a simple technical indicator based on the exponential moving average. Two others, namely Box SVM algorithm and Volume-Weighted SVM, were selected from the literature. Both employ SVM classifier and proved to be effective on other types of capital markets. As it was shown in the empirical analysis, applying machine learning modelling enhances the overall trading performance.

The obtained results are very promising. Assuming achieved scores, the compound return over a one-year period would exceed reasonable levels known from financial markets. It is a direct evidence that Bitcoin market is still young and tends to be ineffective.

Future work will address the issue of position sizing and minimizing the overall risk expressed in this paper by the mean of the maximum drawdown. It is also worth investigating other methods known from traditional markets. One of them can be acquiring data from social networks and use it as an input for machine learning algorithms in order to predict substantial price movements.

References

1. Nakamoto, S.: Bitcoin: a peer-to-peer electronic cash system. Consulted **1**(2012), 28 (2008)
2. Androulaki, E., Karame, G., Roeschlin, M., Scherer, T., Capkun, S.: Evaluating user privacy in bitcoin. In: Sadeghi, A.R. (ed.) Financial Cryptography and Data Security. Lecture Notes in Computer Science, vol. 7859, pp. 34–51. Springer, Berlin (2013)
3. Koshy, P., Koshy, D., McDaniel, P.: An analysis of anonymity in bitcoin using p2p network traffic. In: Christin, N., Safavi-Naini, R. (eds.) Financial Cryptography and Data Security, pp. 469–485. Lecture Notes in Computer Science. Springer, Berlin (2014)
4. Eyal, I., Sirer, E.: Majority is not enough: Bitcoin mining is vulnerable. In: Christin, N., Safavi-Naini, R. (eds.) Financial Cryptography and Data Security, pp. 436–454. Lecture Notes in Computer Science. Springer, Berlin (2014)
5. Żbikowski, K.: Time series forecasting with volume weighted support vector machines. In Kozielski, S., Mrozek, D., Kasprowski, P., Małysiak-Mrozek, B., Kostrzewa, D. (eds.) Beyond Databases, Architectures, and Structures. Communications in Computer and Information Science. Springer International Publishing, Berlin, Vol. 424, pp. 250–258 (2014)
6. Cortes, C., Vapnik, V.: Support-vector networks. Mach. learn. **20**(3), 273–297 (1995)
7. Wen, Q., Yang, Z., Song, Y., Jia, P.: Automatic stock decision support system based on box theory and svm algorithm. Expert Syst. Appl. **37**(2), 1015–1022 (2010)
8. Żbikowski, K.: Using volume weighted support vector machines with walk forward testing and feature selection for the purpose of creating stock trading strategy. Expert Syst. Appl. **42**(4), 1797–1805 (2015)

Part V
Complex Systems, Internet of Things and Agent Systems

Maximal Discernibility Discretization of Attributes—A FPGA Approach

Maciej Kopczynski, Tomasz Grzes and Jaroslaw Stepaniuk

Abstract In this paper we propose the design for hardware cuts generating module for FPGA. Calculations are supported by softcore CPU. Presented architecture has been simulated and tested in VHDL IDE on real data. Implemented algorithm uses Maximal Discernibility (MD) approach. Results show the big acceleration of the computation time using hardware supporting discretization in comparison to pure software implementation.

Keywords Rough sets · FPGA · Hardware · Discretization

1 Introduction

Most real-world datasets consist of both discreet and continuous attributes. This type of data is problematic in case of further analysis, especially if classification system is built. That is why such attributes values must be transformed into discreet form having as small number of values as possible. These transformations must also keep maximum information. In case of rough sets we are trying to preserve discernibility between the objects.

There exist many software implementations of discretization algorithms. However, they require significant amount of resources of a computer system, especially during processing large amount of data.

Field Programmable Gate Arrays (FPGAs) are a group of integrated circuits, whose functionality is defined by the the user using a hardware description language,

M. Kopczynski · T. Grzes (✉) · J. Stepaniuk
Faculty of Computer Science, Bialystok University of Technology,
Wiejska 45A, 15-351 Bialystok, Poland
e-mail: t.grzes@pb.edu.pl
URL: http://www.wi.pb.edu.pl

M. Kopczynski
e-mail: m.kopczynski@pb.edu.pl

J. Stepaniuk
e-mail: j.stepaniuk@pb.edu.pl

D. Ryżko et al. (eds.), *Machine Intelligence and Big Data in Industry*,
Studies in Big Data 19, DOI 10.1007/978-3-319-30315-4_15

such as VHDL. It allows to speed up describing functional properties of the digital system [2]. All these features make a hardware implementation of discretization method in FPGAs possible.

There are some general concepts of the hardware rough set processor, described in [5, 7, 9, 10] as well as the hardware implementations of specific rough set methods, like core evaluation [4, 6], and reduct evaluation [11]. At the moment there is no comprehensive hardware implementation of discretization methods.

Overview of discretization algorithm types, as well as proposition of taxonomy for discretization methods are presented in [3]. The most popular approaches to discretization are EqualWidth, EqualFrequency, MDLP, ID3, ChiMerge, 1R, D2, and Chi2.

The paper is organized as follows. In Sect. 2 some information about the notion of discretization and datasets used during research are provided. The Sect. 3 focuses on description of hardware solution, while Sect. 4 is devoted to the experimental results.

2 Introductory Information

2.1 Discretization of Attributes

Many real-world applications based on rough set methodology involve continuous attributes. However, many of the existing rough set systems cannot handle such attributes directly. In this paper, data discretization is defined as a process of converting continuous condition attribute values into a finite set of intervals without loss of discernibility. Discretization significantly reduces the computation time of various decision rule discovery methods and classification algorithms.

Below one can find definitions and notions used in pseudocode presented in next section. Introductory informations are based on [1].

Let $DT = (U, A \cup \{d\})$ be a decision table, where U is a non-empty finite set of objects, A is a non-empty finite set of condition attributes such that $a : U \rightarrow V_a$ for every $a \in A$. V_a is called the value set of a and V_a is a subset of the set of real numbers and $d \notin A$ is a decision attribute.

An arbitrary attribute $a \in A$ defines a sequence $v_1^a < v_2^a < \cdots < v_{n_a}^a$ of values, where $V_a = \{v_1^a, v_2^a, \ldots, v_{n_a}^a\}$. Then the set of all possible cuts on a is denoted by C_a. The set C_A of all possible cuts on all condition attributes is denoted by $C_A = \bigcup_{a \in A} C_a$.

The main points employed in the Maximal Discernibility Algorithm are to choose the cut $(a, c) \in C_A$ which discerns the largest number of pairs $(x, y) \in U \times U$ of objects from different decision classes $(d(x) \neq d(y))$.

The set C_a of all possible cuts for single attribute a can be calculated using following formula:

$$C_a = \left\{ \left(a, \frac{v_1^a + v_2^a}{2}\right), \left(a, \frac{v_2^a + v_3^a}{2}\right), \ldots, \left(a, \frac{v_{n_a-1}^a + v_{n_a}^a}{2}\right) \right\}. \quad (1)$$

Number of pairs of objects discerned by a cut (a, c) is denoted as $W^X(a, c)$ and is obtained by:

$$W^X(a, c) = l^X(a, c) \cdot r^X(a, c) - \sum_{i=1}^{r} l_i^X(a, c) \cdot r_i^X(a, c) \tag{2}$$

where:

1. $l^X(a, c) = \sum_{i=1}^{r} l_i^X(a, c)$
2. $r^X(a, c) = \sum_{i=1}^{r} r_i^X(a, c)$
3. $l_i^X(a, c) = card\{x \in X : [a(x) < c] \wedge [d(x) = i]\}$
4. $r_i^X(a, c) = card\{x \in X : [a(x) > c] \wedge [d(x) = i]\}$

X is a subset of U, which contains all the objects in input data set. d is the decision attribute. r defines number of of all decision classes in data set.

Number of pairs of objects discerned by a cut $c \notin D$, but not discerned by cuts from D is equal to:

$$W_D(a, c) = W^{X_1}(a, c) + W^{X_2}(a, c) + \cdots + W^{X_m}(a, c) \tag{3}$$

where D is the subset of all possible cuts on all attributes.

Algorithm details will be described in the next section. More introductory information can be found in [1].

2.2 Pseudocode for Generating Cuts

Let $DT = (U, A \cup \{d\})$ be a decision table, where U is a set of objects, A is a set of condition attributes and d is a decision attribute. All condition attributes are real values. Decision attribute consists of discreet values. Below one can find pseudocode based on an algorithm of cuts calculation using Maximal Discernibility (MD) approach, described in [1] as Global Strategy.

INPUT: decision table $DT = (U, A \cup \{d\})$
OUTPUT: semi-minimal set of cuts D

```
1: D ← Ø
2: L ← {U}
3: C ← set of all possible cuts on A
4: repeat
5:   W_Dmax ← 0
6:   (a_max, c_max) ← NULL
7:   for (a, c) ∈ C do
8:     W_D ← 0
9:     for X ∈ L do
10:      calculate W_X for (a, c)
```

11: $W_D \leftarrow W_D + W_X$
12: **end for**
13: **if** $W_D > W_{Dmax}$ **then**
14: $W_{Dmax} \leftarrow W_D$
15: $(a_{max}, c_{max}) \leftarrow (a, c)$
16: **end if**
17: **end for**
18: $D \leftarrow D \cup \{(a_{max}, c_{max})\}$
19: $C \leftarrow C \setminus \{(a_{max}, c_{max})\}$
20: **for** $X \in L$ **do**
21: **if** (a_{max}, c_{max}) divides the set X into X_1 and X_2 **then**
22: $L \leftarrow L \setminus X$
23: $L \leftarrow L \cup \{X_1, X_2\}$
24: **end if**
25: **if** X consists of objects from one decision class **then**
26: $L \leftarrow L \setminus X$
27: **end if**
28: **end for**
29: **until** $card(L) = 0$

Input for the algorithm is the decision table DT. Output is semi-minimal set of cuts D. Lines 1–3 are for initialization of variables. All possible cuts are generated in line 3 using Eq. 1. Main loop in lines 4–29 iterates while set L has any elements in it. L consists of sets partitioned by cuts in D. For loop in lines 7–17 is to find cut that discerns maximum pairs of objects. This loop iterates over all cuts defined on all attributes stored in C. Basing on Eqs. 2 and 3 values W_X and W_D are calculated using loop in lines 9–12. If current W_D value is bigger than maximum W_{Dmax}, then W_D and (a, c) are stored as maximum. In lines 18 and 19 current best cut (a_{max}, c_{max}) is added to output set D and is removed from set C. Last part of algorithm is for recalculating L set. For loop in lines 20–28 iterates over all sets in L. If best cut (a_{max}, c_{max}) divides set X into two parts X_1 and X_2, then set X is removed and X_1 with X_2 are added to L. These operations are performed in lines 21–24. Last part in lines 25–27 is to remove from L sets which contain objects from single decision class.

2.3 Data to Conduct Experimental Research

In this paper, we present the results of the conducted experiments using two datasets downloaded from UCI Machine Learning Repository [8]. First of them is *Iris*. The dataset contains 3 classes of 50 instances each, where each class refers to a type of iris plant. One class is linearly separable from the other 2. The latter are not linearly separable from each other. Dataset consists of 4 condition attributes describing measured features of iris plants (all have numerical values) and 1 decision attribute

Table 1 Example of data encoding for first three objects in Iris dataset

Sepallength	Sepalwidth	Petallength	Petalwidth	Class
5.1	3.5	1.4	0.2	Iris-setosa
4.9	3.0	1.4	0.2	Iris-setosa
4.7	3.2	1.3	0.2	Iris-setosa
0110011	0100011	0001110	0000010	0000
0110001	0011110	0001110	0000010	0000
0101111	0100000	0001101	0000010	0000

(discreet) containing class of iris plant. Second one is *Concerete slump* [12]. The dataset includes 103 objects. It has 10 attributes with numerical values. First 7 attributes describe weight of basic components in 1 m^3 of concrete, while next 3 attributes contain information about the measurement results in given sample of concrete.

Smaller datasets were created as a subsets of original dataset. For the Iris dataset all of them had equal number of objects with given iris class. In case of Concrete slump, smaller datasets were created by choosing given number of objects starting from the beginning of the set.

For the Iris dataset, numerical values on each attribute were encoded as an integer by multiplying real value by 10 and stored using 7 bits for each attribute. Decision attribute value is encoded using 4 bits. Thus every single object is described using 32 bits. Exemplary transformation is presented in Table 1.

In case of Concrete slump dataset, numerical values were encoded also as integer but by multiplying input value by 100. Each value was encoded using 17 bits, what gives 170 bits for all attributes. To ease to fit to memory boundary, single object is described using 192 bits. Unused bits are masked.

3 Hardware Implementation

Hardware solution created by the authors is an implementation of the algorithm presented in Sect. 2.2 for calculating the number of pairs of objects discerned by the given cut. Diagram of the device is shown on Fig. 1. Module consists of the blocks listed below:

- **Decision Table (DT)**—a memory block that stores the decision table objects; every cell (row) contains one object; *Attribute and Cuts Separator Block* must know the structure of every element from decision table;
- **CUTS**—register for storing the cuts for each conditional attribute; structure of this register is the same as *DT* cell structure;
- **MUX (Multiplexer)**—multiplexer for selecting one cell (row) from *Decision Table*; elements are indexed by the *Counter* value;

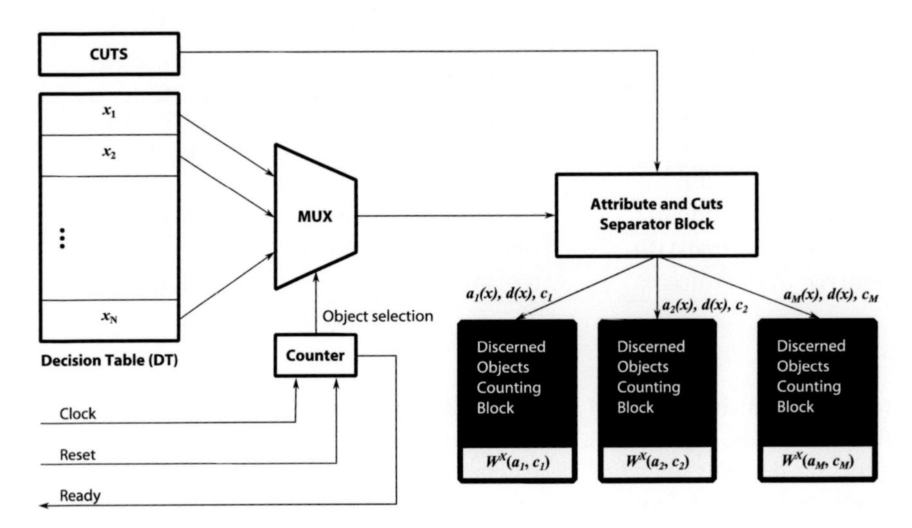

Fig. 1 Diagram of discernibility calculation block

- **Counter**—counter for indexing the elements from a decision table. *Counter* is controlled by external signals: *Clock* and *Reset*. It also provides *Ready* signal for notifying the end of calculations;
- **Attribute and Cuts Separator Block**—block that separates condition and decision attributes values for currently selected object; it also separates the cuts from *CUTS* register;
- **Discerned Objects Counting Block**—block for evaluating the $W^X(a, c)$—number of pairs of objects from X discerned by cut (a, c).

Discerned Objects Counting Block is a block for evaluating the $W^X(a, c)$ using the algorithm described in Sect. 2.2. Diagram of this block is shown on Fig. 2.

- **cl_1, cl_2, ..., cl_K**—counters for calculating the number of objects on the left side of the given cut for each K. K denotes decision attribute value;
- **cr_1, cr_2, ..., cr_K**—counters for calculating the number of objects on the right side of the given cut for each K. K denotes decision attribute value;
- **lr_1, lr_2, ..., lr_K**—registers for storing the value of products $lr\_i = cl\_i \times cr\_i$ for $i = 1 \dots K$;
- **sum_l, sum_r, sum_lr**—registers for storing the sums of the values from registers cl_i, cr_i and lr_i respectively;
- **sum**—register for storing the value of product $sum = sum\_l \times sum\_r$;
- $W^X(a, c)$—output register with the number of pairs of objects from X discerned by cut (a, c).

It must be noticed that presented hardware module is not a standalone cuts calculation solution. It must be supported by proper software which goal is to:

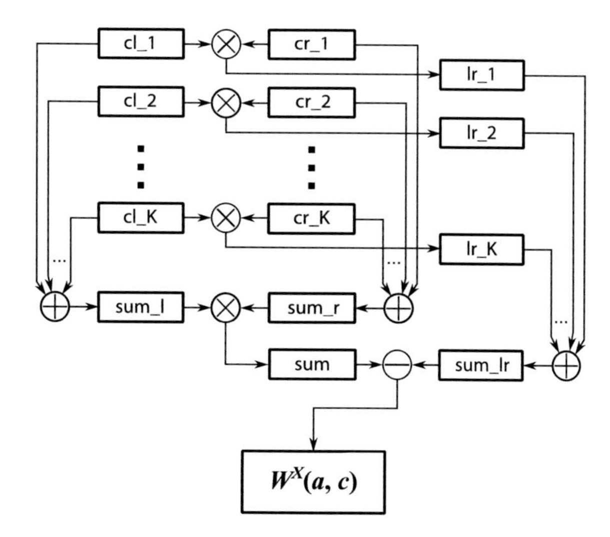

Fig. 2 Diagram of *Discerned Objects Counting Block*

- read and write data to module;
- prepare data from input dataset to transformed form as it was presented in Sect. 2.3;
- sort data;
- perform operations on sets implemented as linked lists;
- control overall operation;

This type of control software has been implemented using NIOS II softcore processor, that can be instantiated inside FPGA chip.

4 Experimental Results

For the research purpose the MD cuts generation algorithm was implemented in C language. Section 2.2 contains details of this method.

The results of the software implementation were obtained using a PC equipped with an 8 GB RAM and 4-core Intel Core i7 3632QM with maximum 3.2 GHz in Turbo mode clock speed, running Windows 7 Professional operational system. The source code of application was compiled using the GNU GCC 4.8.1 compiler. Given times are averaged for 10 000 runs of algorithm with the same set of data.

Quartus II 13.1 was used for design, compilation, synthesis and simulation of the hardware implementation in VHDL language. Synthesis was done for Stratix III EP3SL150F1152C2N FPGA chip. It was assumed, that FPGA clock is running at 50 MHz for the sequential parts of the project, what is 64 times slower than PCs clock.

NIOS II softcore processor, as well as most parts of embedded system were instantiated using Qsys 13.1 tool. Software for NIOS II was implemented in C language using NIOS II Software Build Tools for Eclipse IDE.

Table 2 Comparison of execution time between hardware and software implementation for Iris dataset

Objects	Hardware—t_H (ms)	Software—t_S (ms)	$C = \frac{t_S}{t_H}$	$C_{clk} = 64\frac{t_S}{t_H}$
15	0.021	0.173	8.238	527.238
30	0.087	0.928	10.667	682.667
45	0.198	2.095	10.581	677.172
60	0.354	4.258	12.028	769.808
75	0.555	7.674	13.827	884.930
90	0.801	11.189	13.969	894.002
105	1.092	15.204	14.839	891.077
120	1.428	19.959	13.977	894.521
135	1.809	24.529	13.559	867.803
150	2.235	32.774	13.322	938.495

Table 3 Comparison of execution time between hardware and software implementation for Concrete slump dataset

Objects	Hardware—t_H (ms)	Software—t_S (ms)	$C = \frac{t_S}{t_H}$	$C_{clk} = 64\frac{t_S}{t_H}$
15	0.139	2.970	21.429	1371.429
30	0.574	15.200	26.472	1694.183
45	1.307	41.210	31.535	2018.243
60	2.336	87.100	37.280	2385.893
75	3.663	134.000	36.582	2341.250
90	5.287	212.430	40.183	2571.694
103	6.934	308.000	44.419	2842.820

Tables 2 and 3 present the results of the time elapsed for hardware and software solution. Hardware times were derived from simulation. All times were measured from the moment when dataset was loaded into the memory until obtaining the final result. Last two columns describe the speed-up factor without (C) and with (C_{clk}) taking clock speed difference between PC and FPGA into consideration.

FPGA resources utilization is constant for the different sizes of given input dataset. For different datasets resources utilization varies because of number of bits required to describe single object. Hardware module was compiled to be capable of processing 150 (Iris) and 103 (Concrete slump) objects at a time. Smaller datasets are processed by masking unused objects in the input register. Module uses 3 091 (for Iris) and 31 467 (for Concrete slump) of 113 600 Logical Elements (LEs) total available.

Figure 3 presents a graphs showing the relationship between the number of objects and execution time for hardware and software solution for both datasets.

Presented results show big increase in the speed of data processing. Hardware module execution time compared to the software implementation is 8 (Iris) to 44

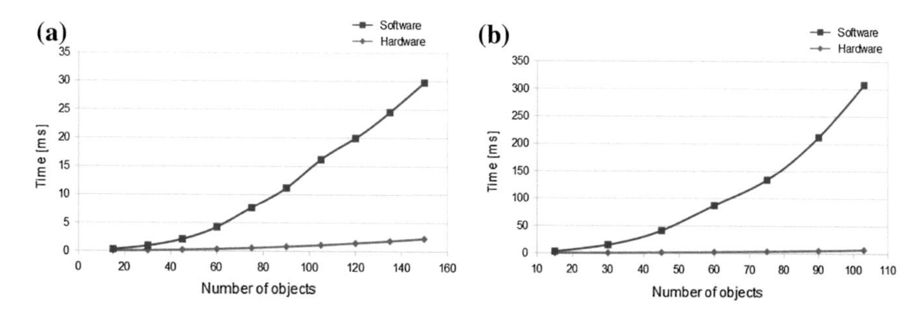

Fig. 3 Relation between number of objects and calculation time for hardware and software implementation for Iris (**a**) and Concrete slump (**b**) datasets

(Concrete slump) times faster. If we take clock speed difference between PC and FPGA under consideration, these results are much better—maximum speed-up factor is 527 (Iris) to 2842 (Concrete slump). Further optimizations give a big chance to increase module calculation time performance.

It can be noticed, that for Concrete slump dataset speed-up factors are bigger than for the Iris and are increasing with number of objects. The reason is that Concrete slump dataset contains more attributes than Iris dataset and the diversity of values are bigger in Concrete slump.

Let comparison of attributes value to given cut or retrieving the element from decision table be an elementary operation. Let assume, that all condition attributes are continuous. k denotes number of conditional attributes, n is the number of objects in decision table, $|L|$ is the number of all objects in sets stored in L, while $card(L)$ is the number of sets in L. $|L|$ is determined by the cardinality of D semi-minimal set of cuts. Computational complexity of software implementation for the MD algorithm is $\Theta(card(L)|L|nk)$. $|L|$ is equal to n at beginning of algorithm, while $card(L)$ is 1, so we can write that complexity is $\Theta(n^2k)$. Using hardware implementation, final complexity of cuts generation is $\Theta(n^2)$. The k is missing, because our solution performs comparison between all attributes in $\Theta(1)$. For single object, all attributes values for given cut are compared in single clock cycle. In most cases $k \ll n$, so we can say, that computational complexity for software and hardware implementations have the same order.

5 Conclusions and Future Research

The hardware implementations of discretization methods are very competitive to high computing power PCs, especially in real time solutions. As it was presented, performing cuts generation using hardware implementation gives us a big acceleration in comparison to software solution.

It should be noticed that hardware implementation does not offer such flexibility as software versions of algorithms. Hardware module must be recompiled for different structures of datasets (eg. attributes number change). But the biggest advantage of FPGA is possibility of its reprogramming.

Further research will focus on scaling presented solution to much bigger datasets. Very important thing is to create implementation of hardware supported sorting algorithm, such as merge sort. Another improvement in calculation time can be achieved by introducing second level of parallelism by multiplying hardware modules processing subsequent parts of dataset. First level was implemented by comparing given cut with all attributes values for single object.

Acknowledgments The research is supported by the Polish National Science Centre under the grant 2012/07/B/ST6/01504 (Jaroslaw Stepaniuk, Tomasz Grzes) and by the scientific grant S/WI/3/2013 (Maciej Kopczynski).

References

1. Bazan, J.G., Nguyen, H.S., Nguyen, S.H., Synak, P., Wróblewski, J.: Rough set algorithms in classification problem. In: Polkowski, L., Tsumoto, S., Lin, T.Y. (eds.) Rough Set Methods and Applications, Studies in Fuzziness and Soft Computing, vol. 56, pp. 49–88. Physica-Verlag, Heidelberg (2000)
2. Bezerra, E., Lettnin, D.V.: Synthesizable VHDL Design for FPGAs. Springer, New York (2014)
3. Garcia, S., Luengo, J., Sáez, J.A., López, V., Herrera, F.: A survey of discretization techniques: taxonomy and empirical analysis in supervised learning. In: IEEE Transactions on Knowledge and Data Engineering, vol. 25, issue 4, pp. 734–750. IEEE (2013)
4. Grześ, T., Kopczyński, M., Stepaniuk, J.: FPGA in rough set based core and reduct computation. In: Lingras, P., Wolski, M., Cornelis, C., Mitra, S., Wasilewski, P. (eds.) Rough Sets and Knowledge Technology: 8th International Conference: RSKT'2013. Lecture Notes in Computer Science, Lecture Notes in Artificial Intelligence, vol. 8171, pp. 263–270. Springer, Berlin (2013)
5. Kanasugi, A., Yokoyama, A.: A basic design for rough set processor. In: The 15th Annual Conference of Japanese Society for Artificial Intelligence (2001)
6. Kopczyński, M., Grześ, T., Stepaniuk, J.: Generating core in rough set theory: design and implementation on FPGA. In: Kryszkiewicz, M., Cornelis, C., Ciucci, D., Medina-Moreno, J., Motoda, H., Raś, Z.W. (eds.) Rough Sets and Intelligent Systems Paradigms: Second International Conference: RSEISP'2014: Held as Part of JRS. Lecture Notes in Computer Science, Lecture Notes in Artificial Intelligence, vol. 8537, pp. 209–216. Springer, Berlin (2014)
7. Lewis, T., Perkowski, M., Jozwiak, L.: Learning in hardware: architecture and implementation of an FPGA-based rough set machine. In: 25th Euromicro Conference (EUROMICRO'99), vol. 1, pp. 1326 (1999)
8. Lichman, M.: UCI machine learning repository. University of California, School of Information and Computer Science. Irvine. http://archive.ics.uci.edu/ml (2013)
9. Pawlak, Z.: Elementary rough set granules: toward a rough set processor. In: Pal, S.K., Polkowski, L., Skowron, A. (eds.) Rough-Neurocomputing: Techniques for Computing with Words, Cognitive Technologies, pp. 5–14. Springer, Berlin (2004)
10. Stepaniuk, J., Kopczyński, M., Grześ, T.: The first step toward processor for rough set methods. Fundam. Inform. **127**, 429–443 (2013)
11. Tiwari, K.S., Kothari, A.G., Keskar, A.G.: Reduct generation from binary discernibility matrix: an hardware approach. Int. J. Futur. Comput. Commun. **1**(3), 270–272 (2012)
12. Yeh, I.-C.: Modeling slump flow of concrete using second-order regressions and artificial neural networks. Cem. Concr. Compos. **29**(6), 474–480 (2007)

Big Data Solutions for Smart Grids and Smart Meters

Joanna Konopko

Abstract The article describes the architecture of Big Data systems, explains its architecture and also clarifies what Big Data is. The document also presents basic problems related to management Big Data in Smart Grids infrastructure and smart meters. Advanced Metering Infrastructure is described in the example of current implementation status, plans and perspectives on pilot projects already carried out by the PGE Dystrybucja SA in two locations: Łódź City and Augustów City. The current state of data generation of the Polish grid is analyzed and a future realistic scenario is illustrated.

Keywords Big data · Smart grids · Smart meters

1 Introduction

Living in the information age, watching the progress of technology, we can conclude that, information has become the new natural resource, as valuable as mineral resources. The control of information resources and their processing is an essential element of domination and social and economic majority. An example would be the economy of the United States, which has launched a program run by the National Security Agency with the aim of telephone data collection, and data from the servers of Internet companies of its citizens. The corporation Booz Allen Hamilton, whose task was commissioned, earned on this program last year alone $1.3 billion. As reported by The New York Times [1] as a lucrative contract, amounting to the $5.6 billion was signed for the next five years. This has clearly confirmed the importance of information nowadays.

However, the data is not sufficient. As a result of progress and visible changes in society, it is becoming easier to access the media. The access to the Internet is becoming a universal service, and according to a report from Cisco by the end of

J. Konopko (✉)
Institute of Computer Science, Warsaw University of Technology,
Nowowiejska 15/19, 00-665 Warsaw, Poland
e-mail: j.konopko@ii.pw.edu.pl

D. Ryżko et al. (eds.), *Machine Intelligence and Big Data in Industry*,
Studies in Big Data 19, DOI 10.1007/978-3-319-30315-4_16

2017 half the people will have access to it [2]. The report predicts the number of 3.6 billion internet users, which will represent more than 48 % of the world population, the projected world population of 7.6 billion. It is also estimated tripling of global IP traffic from 2012 to 2017. Global IP traffic will reach 1.4 zettabytes per year in 2017, or 120.6 exabytes per month. (1 zettabyte = 1000^7 bytes = 10^{21} bytes = 1000 exabytes = 1 billion terabytes.). These projections define a tendency to seek and use the hidden value in the available volumes of data.

The growth of data is observed in every area of life [3]: medicine, tourism industry, banking, finance, insurance, e-commerce, etc. The data is collected in an increasingly faster rate and its analysis takes more time.

The LHC Computing Grid is an international collaborative project that consists of a grid-based computer network infrastructure connecting 140 computing centers in 35 countries. It was designed by CERN to handle the significant volume of data produced by LHC experiments. By 2012 data from over 300 trillion (3×10^{14}) LHC proton-proton collisions had been analyzed, LHC collision data was being produced at approximately 25 petabytes per year, and the LHC Computing Grid had become the world's largest computing grid (as of 2012), comprising over 170 computing facilities in a worldwide network across 36 countries [4].

The Sloan Digital Sky Survey (SDSS) [5] is one of the most ambitious and influential surveys in the history of astronomy. Over eight years of operations it obtained deep, multi-color images covering more than a quarter of the sky and created 3-dimensional maps containing more than 930 000 galaxies and more than 120 000 quasars. When the Sloan Digital Sky Survey (SDSS) began collecting astronomical data in 2000, it amassed more data during first few weeks than it was collected in the whole history of astronomy at that time. Continuing at a rate of about 200 GB per night, SDSS has amassed more than 140 terabytes of information. When the Large Synoptic Survey Telescope, successor to SDSS, comes online in 2016 it is anticipated to acquire that amount of data every five days. Decoding the human genome originally took 10 years to process, now it can be achieved in less than a week [6]. Facebook handles 50 billion photos from its user base. Another example is an American multinational retail corporation Walmart, which handles more than 1 million customer transactions every hour, which are imported into databases estimated to contain more than 2.5 petabytes (2560 terabytes) of data. The NASA Center for Climate Simulation (NCCS) stores 32 petabytes of climate observations and simulations [7].

However, the faster process of collecting data, causes a direct increase of the size of the database. Unfortunately, the progress is not in line with the increasing the ability to analyze the acquired data.

For years, we have witnessed the evolution of database systems, from simple file systems to relational and object-oriented systems and data warehouses. Particularly noticeable is the process of competing with analytics, in other words, competing on analytics [8].

Companies take such actions as predicting user behavior and optimizing the related marketing activities, especially chains, logistics, services and products to achieve a significant advantage in the market and increase their competitiveness.

Currently analyzed unstructured data, an example is called "text mining", which will significantly expand the scope of knowledge necessary to make the right business decisions. However, when analyzed the content of an increasing volume of data and they come from many different places, they are constantly updated and is still growing, the analytical capabilities and potential use of the knowledge stored in these data is growing exponentially. The data is often generated for its countless sources, often from outside of the organization, from the areas so far completely unqualified as a source of information relevant to making the right business decisions. This determines the research, including coverage innovative technological tools and innovative sets of competences.

According to the report [9] made by IDC Digital Universe commissioned by EMC Corporation, the size of the digital universe by 2020 will 40ZB so 14 % more than expected. 5247GB (gigabytes) of data will fall per capital in the world. Expenditure on infrastructure and IT solutions, software, services, telecommunications and information technology staff will increase in the years 2012–2020 by 40 %. According to IDC forecasts, the market for Big Data will grow from 3.2 billion dollars in 2010 to 16.9 billion in 2015. Statement also reveals the vulnerability of large sets of data: less than 1 % of the data is analyzed, and less than 20 % is protected. The main reason for the expansion of the universe of digital data generated by machines, in 2005, they accounted for 11 % of all data, while in 2020 this ratio exceeds 40 %.

All of this indicates the need for new and better analytical tools by which the data analysis process is not only safe but also efficient. It is predicted that the data analyst profession will be one of the most popular professions in twenty-first century [10]. According to the report [11] made by the McKinsey Global Institute, the biggest factor of economic growth by 2020 in the economy of the United States is an area of Big Data, while introducing the notion of an economy based on the data.

2 The Organization of a Paper

This paper deals with the following research questions: *What a Big Data is and whether Smart Grid systems could be an example of Big Data systems?*

The rest of the paper is organized as follows: Sect. 3 briefly defines the Big Data concept and offers some examples of today's sources of data. Section 4 provides an overview about architecture of Big Data system and about the concept of Smart Grid. Section 5 provides a description of examples of implementation pilot projects in Poland. Advanced Metering Infrastructure is described in Sect. 6 on the example of current implementation status, plans and perspectives on pilot projects already carried out by the PGE Dystrybucja SA in two locations: Łódź City and Augustów City. Section 7 provides a quantitative analysis of the amount of data generated by a smart grid infrastructure with special focus on the Polish grid as concrete example. The concluding remarks complete the paper, Sect. 8.

3 Big Data

The concept of Big Data is a new direction, and the first attempt to define this concept took place in 2001. Then the company META Group (now Gartner) published a report [12] which described the Big Data in the 3 V model. The presented scheme describing the large amount of data—volume, high variability in the data, the speed they are received and analysis—velocity and a large variety of data—variety.

Typically, large sets of data (Big Data) contain information about the size beyond the ability of commonly used computational software tools, management and processing of data within a predetermined time range [13]. This means that at times, the content cannot be analyzed correctly for a specific time regime due to too much data volume. The size of Big Data is a variable value, since 2012 are the size of a few hundreds of terabytes to multiple petabytes of data in a single data set.

In 2012, Gartner completed the previously given definition "*Big Data is high volume, high velocity, and/or high variety information assets that require new forms of processing to enable enhanced decision making, insight discovery and process optimization.*" [14]). In addition to the 3 V model sometimes added yet another value: the credibility of data—veracity [15], or data value [16].

To properly define what Big Data is we need to look closely at the three values of the 'V' that describe them.

Volume—the amount of data. The technology is very restrictive, focusing on a specific data size. According to a report by the McKinsey Global Institute [11] the concept of big data refers to data sets whose sizes exceed the capacity of ordinary tools for the collection, storage, management and analysis. This definition is flexible, does not talk about the specific values of the size of the data but the technological capacity to manage them. For some people they may be of the order of terabytes and petabytes to others.

Velocity—the rate of appearance of new data. To extract the important information flowing and rapidly changing content, must be analyzed in near real time. Today in business, speed is important to obtain valuable, critical for enterprise information. A commercial from IBM makes the point that you wouldn't cross the road if all you had was a five-minute old snapshot of traffic location [17].

Variety—variability of the data. Information is constantly flowing from a variety of sources, including transactional systems, web services, social networks, mailboxes, databases. There are different data formats: documents, music, videos, news, forms, website traffic, phone calls, surveys, offline data, and social networking content. These data are unstructured and unsuitable for traditional forms of analysis related to the relational model combining data from the same sources.

Additional value complementary to the model 4V:

Veracity—reliability of the data. The information must be true, it is necessary also their integrity, due to these characteristics organizations are able to trust the data they receive, and after their analysis and processing make key business decisions.

Value—the data value. Given the huge amount of information, a large majority of it is unnecessary, or useless from a business point of view. An important factor is significantly valuable data filtering and omission of useless content. In publications we can also find an additional term which is equivalent in many respects to the idea of "Big Data"—it is the High Performance Analytics, which advanced to analyze large volumes of data.

4 Architecture of Big Data System

Big Data system generally consists of three segments: storage system, processing and analysis [18], presented in Fig. 1. Choosing an architecture and building an appropriate Big Data solution is challenging because various factors (especially those listed previously: volume, velocity, variety, veracity, value) have to be considered. A storage system is responsible for acquiring data from the data sources. The processing segment converts the data to a format that could be analyzed. The analysis segment reads the data digested by the data massaging and storage system. In some cases, the analysis segment accesses the data directly from the data source.

4.1 The Concept of Smart Grid

Figure 2 shows a Smart Grid with distributed power generation from large solar and wind farms, as well as bidirectional flow from producer-customers of various sizes. Only an intelligent grid can integrate such widespread electricity generation from renewable sources, along with outputs from traditional large generating stations, and deliver power in real time to customers with varying requirements. Decentralization,

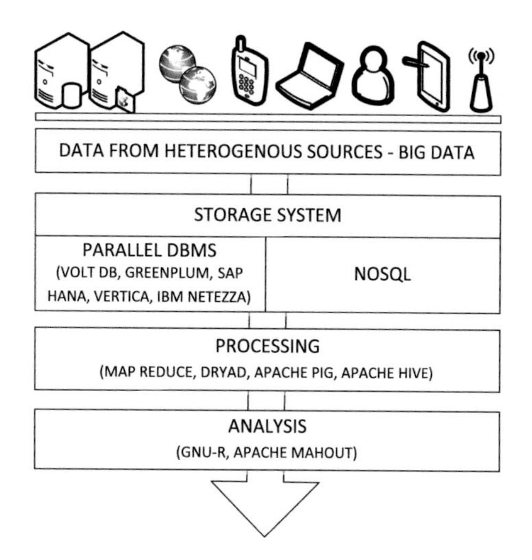

Fig. 1 Architecture of big data system [18]

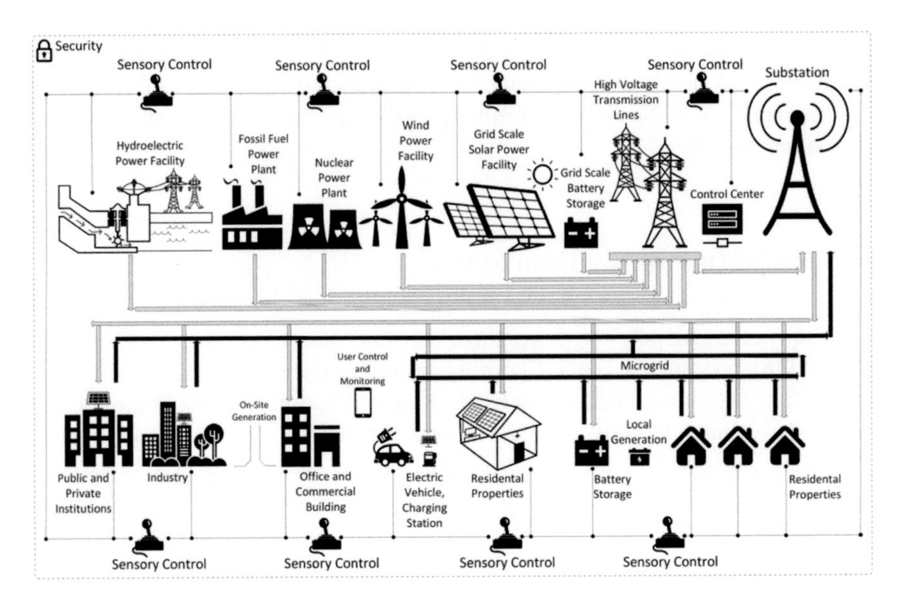

Fig. 2 Smart grid architecture (own work)

in turn, can happen only through smart grids that are enabled by advanced electronic measurement, control and communication. The locally-generated power may be consumed directly by the customer/generator, but, as markets and regulations permit, it may also be sold as excess back to the central utility through the grid or even sold to neighbors through a microgrid. The longstanding paradigm of one-way flow of electricity is succeeded by two-way and multi-way power flow, much like the flow of communications over the Internet today. Sense and control mechanisms will be deployed throughout the system and this will place new demands on operators of generating plants and transmission networks.

This new system of distributed energy generation will provide customers with additional ways to reduce their power costs satisfy their desire for cleaner power choices and manage their energy production to ensure it meets their quality and availability requirements.

4.2 Smart Meter

A smart meter is a digital electric meter that measures and records usage data hourly, or more frequently, and allows for two-way communications between the utility and the customer.

The example of smart meter is shown in Fig. 3. The electricity meter GAMA 100 is developed for residential applications and is approved according to EN 50470-1, EN 50470-3, IEC 62052-11 and IEC 62053-21. Data transmission meets specification of IEC 62056-21, IEC 62056-31,EN 13757-2 and EN 13757-3 protocols. The electricity

Fig. 3 Single-phase static electricity meter GAMA 100 [19]

meter satisfies the requirements of Directive 2004/22/EC of the European Parliament and of the Council of 31 March 2004 on measuring instruments.

With Automated Metering Infrastructure (AMI) technology and a display showing both consumption and price, the customer can reduce consumption when prices are high and get full benefits from it with accurate metering and billing based on the actual consumption. In this way, customers can save on their electricity bills. Retailers can compete by offering customers different electricity prices which apply at different times of the day. Similarly, retailers can target certain groups of customers with particular tariffs that would be most economical for their consumption patterns. Such pricing innovation and variety in smart meters will promote retail competition to the benefit of customers. Especially for households with relatively high electricity consumption, house and building automation can lead to additional savings and increased comfort: optimization of heating and lighting of the household and small business.

5 Smart Grids in Poland

Figure 4 shows the main distribution system operators in Poland. As far as publicly visible initiatives are concerned the following leading actors can be listed: the Polish National Energy Conservation Agency (KAPE)—leader in promoting SG ini-

Fig. 4 Distribution system operators in Poland: Energa, ENEA, PGE, RWE, Tauron [20]

tiatives, involvement IEE Programme—Smart Regions IEE programme; NFOSiGW preparing (National) Priority Programme on Smart Grids, involved in the European Electricity Grids Initiative (EEGI) and the European Walk to Smart Grids Excellence" ERA-NET; Polish Association of Transmission and Distribution of Electrical Energy (PTPiREE)—represents and secures interests of electrical energy sector in some activities, orders analysis of impact of AMI on power system.

Smart meters are playing a critical role in shaping the electric grid of tomorrow and enabling the integration of new technologies and innovations across the grid. As the power grid evolves into a broad platform for integrating new energy services and technologies, the ability to connect legacy assets and systems and integrate new ones is critical; smart meters are supporting this evolution. In addition, the data collected by smart meters (or automated metering infrastructure (AMI)) opens the door for greater integration of new resources and new energy services for customers.

All leading pilot projects on smart meters in the electric power sector were identified and selected examples are shown in Table 1.

6 Pilot Project Carried Out by the PGE Dystrybucja SA

"ELGAMA-ELEKTRONIKA" Ltd. is the leading manufacturer of static electricity meters in the Baltic region. The company won the tender for smart metering system in two pilot locations: Augustów City and Łódź City [23]. Smart Metering Solution from this company is end-to-end solution for utilities to control their energy including metering, data collection and organizing, validation and analysis. Solution comprises all the necessary constituents of an advanced metering infrastructure: meter data

Table 1 Some leading pilot projects on SM identified in Poland (own work on the basis of [21, 22])

Full Project Title	Abstract	Thematic focus of the project
AMI implementation in Energa Operator [3]	Project scope: Implementation of AMI infrastructure	- Integration of Smart Metering
	- Installation of intelligent meters (municipal—2 500 000; business—300 000)	- Metering infrastructure; - Smart metering data processing
	- Installation of balancing meters at LV/MV transformer stations	
	- AMI application (central data repositorium)	
	- ICT layer (PLC, iMAX) Budget: EUR 300 000 000	
	This is the biggest AMI Project in Poland Goals: Remote metering data acquisition, Remote control of metering devices, Increase in operational efficiency, Optimization of grid management and development processes, Increase of customer awareness in the areas of energy efficiency and distributed generation, Accordance with regulatory requirements for meter reading	
Intelligent Hel Peninsula	Goals	- Integration of DER and new users
	- Automation of grid management	- DSO Integration of small DER; Infrastructure to host EV/PHEV; Integration of Smart Customers
	- DER (prosumers, creating possibilities for integration with intelligent homes)	
	- PHEV (grid adaptation, energy storage)	

(continued)

Table 1 (continued)

Full Project Title	Abstract	Thematic focus of the project
AMI–Pilot project	Project scope: Implementation of AMI infrastructure	- Integration of Smart Metering
	- Installation of intelligent meters (approx. 3400)	- Metering infrastructure; Smart metering data processing
	- Installation of balancing meters (PLC/GPRS) Goals	
	- Constant monitoring of supply quality	
	- Remote system management	
	- technology and system functionality verification	
	- preparation for project's next stage (installation of 50 000 smart meters)	
Remote data reading from users in ENEA Operator	Project scope	- Integration of Smart Metering
	- Installation of intelligent meters (approx. 1000)	- Metering infrastructure; Smart metering data processing
	- access points at the transformer stations (PLC/GSM/GPRS)	
	- Installation of balancing meters (PLC/GPRS) Goals	
	- Remote system management	
	- Remote data readings (15 min)	
AMI and Smart Grid Pilot Project	Project scope: Implementation of AMI infrastructure	- Integration of Smart Metering
	- Installation of intelligent meters (Approx 2600 PLC - LV/MV; approx. 2200 RADIO/GPRS)	- Metering infrastructure; Smart metering data processing
	- Automation of MV grid and MV substations	
	Goals: CBA analysis; preparation for project's next stage (installation of 30 000 smart meters)	

management software (MDMS), communication infrastructure and smart metering devices. Open standards based solution is fully scalable and flexible, that allows building bi-directional metering and data collection infrastructure.

6.1 Augustów

See Fig. 5.

6.2 Łódź

See Fig. 6.

Fig. 5 Augustów location (on the basis of [24])

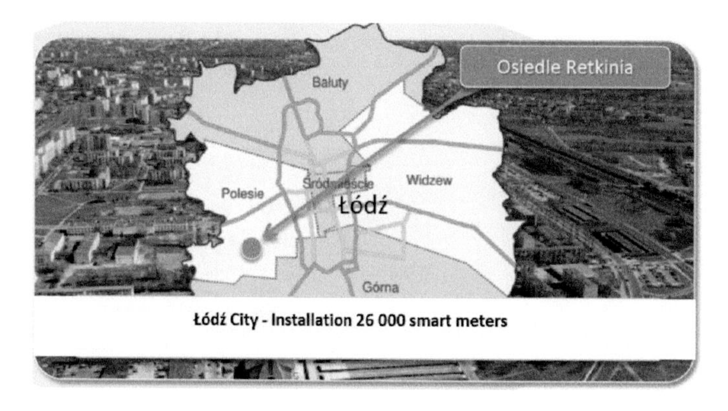

Fig. 6 Łódź location (on the basis of [24])

6.3 Metering Data Management System

Metering Data Management System (MDMS) is designed to organize near real-time accounting of the consumption of power, heat, water, steam, gas and others energy resources. The system was originally created for a big number of meters, facilities and users. The architecture of the software allow scale the system hardware per project base to achieve needed performance. Functionally (for large-scale systems and on a unit-to-unit basis), the software modules are divided into three groups:

1. Database server(s) which runs on Oracle RDBMS for the storage and processing of information. It ensures high productivity of the system and possibility of connecting a great number of meters.

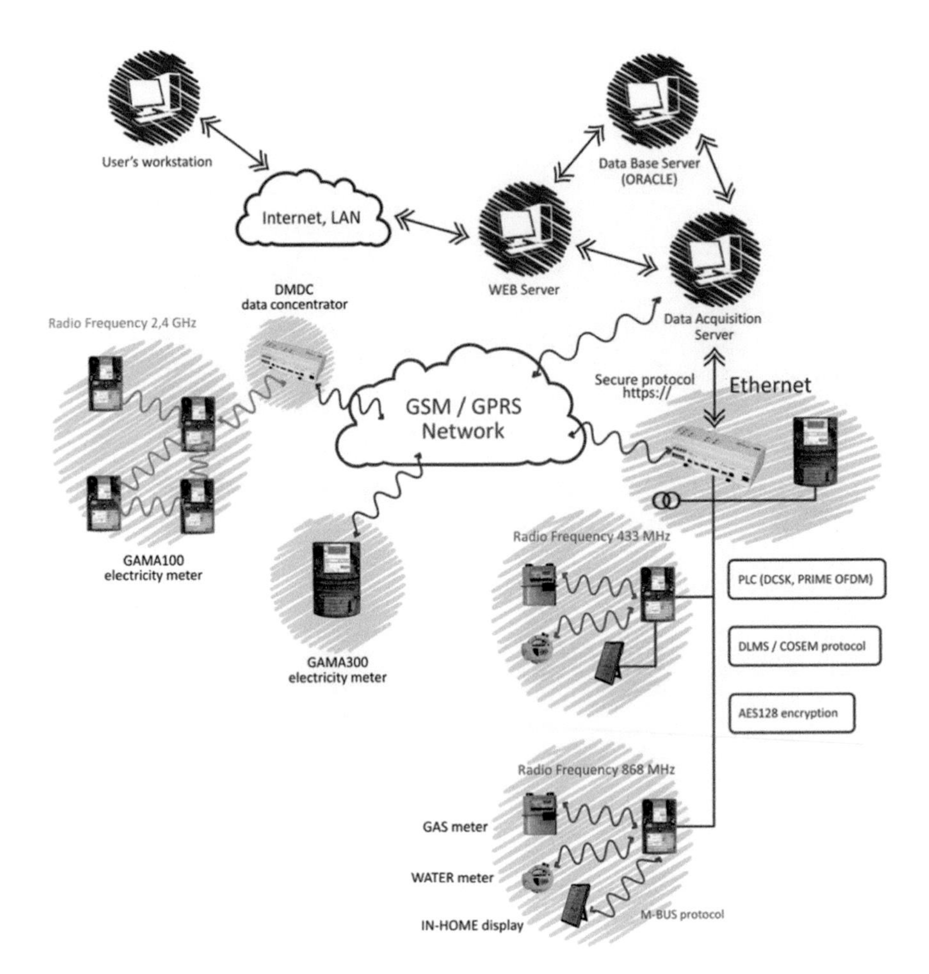

Fig. 7 Smart metering system [25]

2. Data Collection Server(s) ensures collection of data of commercial metering and their placement into the database. It can open hundreds of sessions with facilities in parallel, which allows drastic reduction in the time needed for data collection and switching between alternative communication channels. The architecture allows flexible integration of new devices.
3. WEB server(s) for the provision of information to the Automated Working Station. The user interface is fully based on WEB technologies. The portal allows complete abandonment of specialized software at work places and operation of the system from any computer that has an internet browser (subject to use rights) and access to the network of the portal.

Client Portal allows consumers to access detailed information about their electricity consumption, by providing access to data of the load diagram, allowing a detailed view of consumption for the selected period, and also by providing a comparison with the average consumption of typical electricity consumers, as well as various pieces of advice on energy efficiency (Fig. 7).

Smart Metering System consists of smart meters: GAMA series meters are dedicated to smart grids and designed for secure data communications in open network. Meters can be equipped with various communication modems according to metering point position and available data transfer technology. In terms of the network topology, the communication infrastructure is built on various communication technologies: Ethernet and GSM/GPRS for the wide area network (WAN), PLC and radio-based solutions for last mile communication, wired and wireless technologies for the home area network (HAN). Metering Data Management System (MDMS) is designed to manage all processes of a Smart Metering, including data collecting, network configuration, analysis, validation and other data managing operations.

6.4 Functionality

MDMS can be divided to subsystems by functionality:

- Information Collection subsystem. It performs tasks related to monitoring of schedules for the collection of measuring data, data on the state of facilities and metering devices, repeated queries in case of session failures and recovery of missing data.
- Data verification subsystem. By continuously verifying received data, it is responsible completeness and validity of archived data. It allows keeping track of balances during data collection.
- Monitoring subsystem. The subsystem is logging absolutely all events: failures, detection of incorrect data, communication errors, full auditing of user actions and much more. It enables the real-time delivery of e-mail and SMS messages to interested parties upon the occurrence of events (alarms/alerts).
- Real-time Query Subsystem. It is designed for the real-time query to data concentrators and meters for the purpose of executing test queries. By selecting devices,

preferred parameters and dates, a user can initiate a query from the data collection server to a device and, having received a response, view, print out, export it to Excel.

- Visualization subsystem. It enables to get access to the results of operation of all subsystems. It is possible to work with archives, create groups and view information by groups, view data in the form of symbolic circuits and design preferred reports.
- Safety subsystem. Is designed to keep track of security of data and communication. By performing internal system self-testing and external communication encryption and decryption the global safety of the system is ensured.
- Exchange with external systems. It is designed for the automatic delivery of information on measurement results, states of measuring devices and measurement objects. Flexible integration of new exchange protocols is available. Various formats (XML, CSV ASKP and others) and transports (SOAP, FTP, SMTP/POP/IMAP and others) are implemented. CIM model and data exchange/control (as slave system) according IEC61968-9 also supported.
- Analytical subsystem. The data collected goes through all sorts of crosschecks to identify any inaccurate information, or abnormal consumption or theft. The number of algorithms and checks is limited only by the imagination of the user. The system provides the user with a tool to implement any of these algorithms.

7 Analysis

Taking Poland as a basic case study, the current and a possible future situation for the power grid are considered, specifically considering the amount of generated data. Currently smart meters are only partially installed in the country (as it is shown in Table 1) and the approach is first to test the AMI in controlled settings and in pilots before proceeding to a massive roll out, expected to cover the country by 2020. In our quest for Big Data, the current situation of advancement of the AMI and smart grid in Poland is considered and two realistic scenarios of the near future are presented. The current situation with two scenarios: the near future and the far future scenario are synthesized in Table 2.

Table 2 Polish Smart grids data

	Current Pilot Projects	Near Future Scenario (80 % of consumers up to 2020)	Far future Scenario (100 % of consumers)
	Metering		
Installed smart meters in Poland	1 mln	16 mln	17 mln
Smart meter sampling period	2 h	15 min	5 min

Table 3 Data size for various parameters/devices/services in bytes per sample [28]

Data use in metering	Size (bytes)
Consumption only	193
Consumption/production	245
Consumption/production, instantaneous power (3-phase), and current	530
Consumption/production, instantaneous power (3-phase), current, failures, and gas metering	1100

The 100 % penetration is based on the commitment of the European Union to deploy smart meters to at least 80 % of all customers (c.f. EU directives 2009/72/EC and 2009/73/EC) and the commitment of the Polish government. According to [26, 27] the number of installed smart meters in near future scenario is 16 mln and in far future scenario is 17 mln. Implementation of smart metering solutions is associated with huge amount of generated data, often sampled at very high frequency (hours, minutes and even seconds). In near future scenario the sampling period is 15 min and in far future scenario is 5 min (Table 3).

For the Big Data estimation, four examples are identified. Each case has increasing data quantities as more units are involved. The generated data can be identified from many different sources: Example A: Consumption data from smart metering Example B: Consumption/production data from smart meters Example C: Consumption/production, instantaneous power (3-phase), and current from smart metering. Example D: Data for consumption/feed-in, instantaneous power, failures from smart metering.

In Fig. 8 the yearly data generation from smart meters is presented. The amount of data can be measure using megabytes, so it is not comparable to size of Big Data system examples identified by dozen of terabytes.

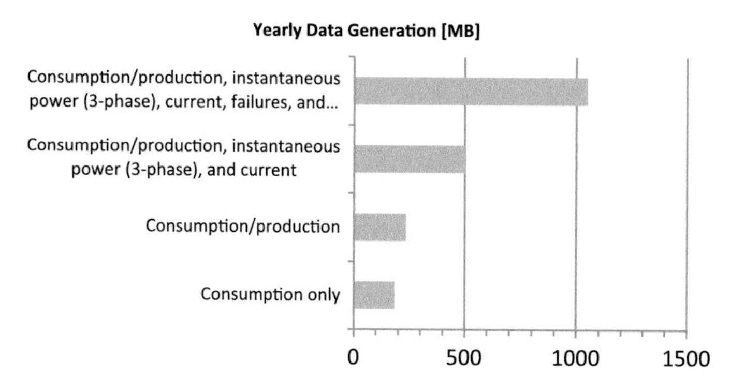

Fig. 8 Current Polish AMI yearly data generation in MB

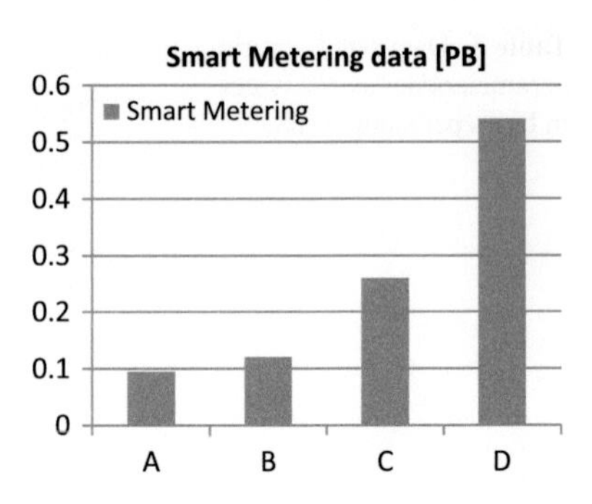

Fig. 9 Polish smart grid yearly data generation in PB: near future scenario

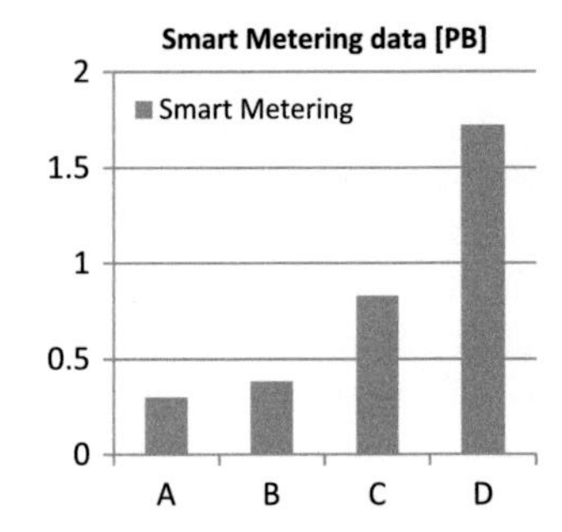

Fig. 10 Polish smart grid yearly data generation in PB: far future scenario

In Fig. 9 the simulation of data generated in near future scenario is presented and in Fig. 10 the data generated in far future scenario is presented. Considering the near future scenario for full smart grid implementation the amount of data produced yearly for the whole of Poland starts to assume the connotation of Big Data. The sampling frequency is 15 min for metering. Considering the far future scenario, with even higher penetration of smart devices and higher sampling rates, the amount of data naturally grows. The sampling period of 5 min for metering-related data. In Example D one reaches the level of about 1,7 petabytes.

One aspect that is not considered in simulation—the fact that all the data containing user and privacy related information will need to be protected by an additional layer of security (e.g., encryption) that require even more bits to be transmitted and stored.

In Table 4 the data use in smart devices is provided and also a data use in weather forecast is presented. The number of electric vehicle is estimated on based on [29], in far future scenario 50 % of population will have a car and in near future one per 10 passengers will have a car. 90 000 battery packs is estimated for local energy storage, and 180 000 in far scenario (Table 5).

In Fig. 11 the simulation of data generated in near future scenario is presented and in Fig. 12 the data generated in far future scenario is presented. Considering the near future scenario for full smart grid implementation the amount of data produced yearly for the whole of Poland starts to assume the connotation of Big Data. The sampling frequency is 15 min for metering, electric vehicles and batteries (total measures in one year is 34560) whereas the weather information are provided every hour (total measures in one year is 8640). It is note that more than half of the data

Table 4 Data size for various parameters/devices/services in bytes per sample [28]

Data use in smart devices	Size (bytes)
Electric vehicle consumption only	193
Electric vehicle consumption/feed-in	245
Electric vehicle consumption/feed-in, instantaneous power (3-phase), and current	530
Electric vehicle consumption/feed-in, instantaneous power (3-phase), current, failures, and gas metering	1100
Battery	200
Data use in weather forecast	Size (bytes)
Essential weather parameter	13000
Improved weather parameter	15000
Advanced weather parameter	20000

Table 5 Number of smart devices in Poland

Smart devices			
Electric vehicles	70	3853329	19266649
Battery packs	N/A	90 000	180 000
Intelligent appliances per household	0	10	20

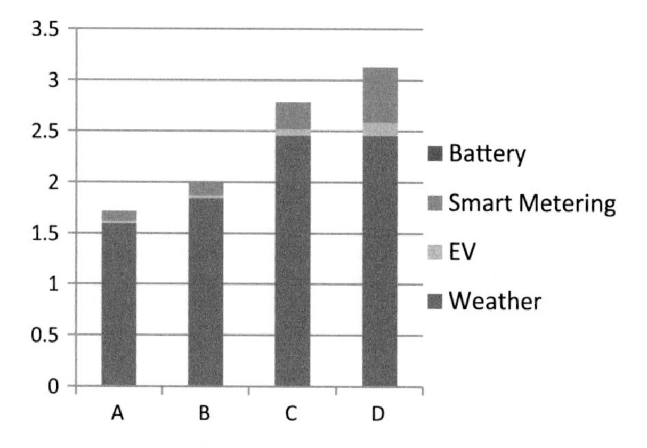

Fig. 11 Polish smart grid yearly data generation in PB near future scenario

generated comes from the meteorological information. Considering the far future scenario, with even higher penetration of smart devices and higher sampling rates, the amount of data naturally grows. The sampling period of 5 min for metering-related data, and 30 min for weather. In Example D one reaches the level of about 9 petabytes, distributed between weather information (5 petabytes), electric vehicle information (about 2 petabytes), and smart meters (1,7 petabytes). One aspect that is not considered in simulation—the fact that all the data containing user and privacy

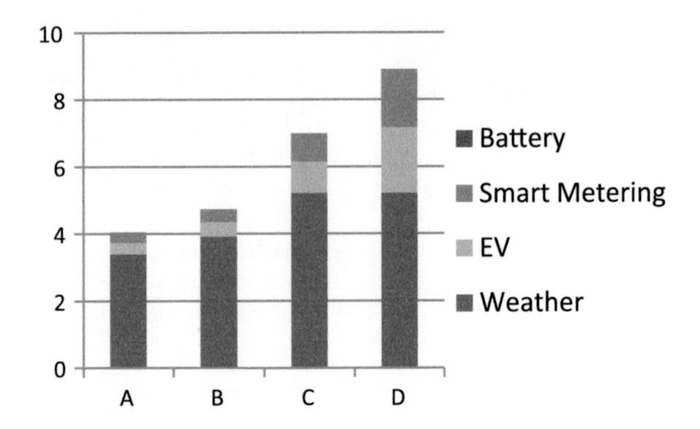

Fig. 12 Polish smart grid yearly data generation in PB far future scenario

related information will need to be protected by an additional layer of security (e.g., encryption) that require even more bits to be transmitted and stored. On the other hand, any compression level possible for the smart grid data is not considered, but the figures here presented are raw data.

8 Conclusion

Given the current state of affairs, the smart grid is not comparable with the examples of Big Data systems. However, when more and more houses and businesses will be equipped with smart meters then the Polish smart grid will become closer to be a Big Data system. In this paper, projections and calculations were made on how the relatively small grid of Poland could become a Big Data generator similar to Facebook. Interesting challenges are not only in the management of the future smart grid data, but also on how to extract value from this data and on novel business models based on the availability of Big Data.

Acknowledgments The study is co-financed by the European Union under the European Social Fund. Project PO KL "Information technologies: Research and their interdisciplinary applications", Task 6, agreement No UDA-POKL.04.01.01-00-051/10-00.

References

1. Appelbaum, B., Lipton, E.: Leaker's Employer Is Paid to Maintain Government Secrets. The New York Times. http://www.nytimes.com/2013/06/10/us/booz-allen-grew-rich-on-government-contracts.html?_r=0
2. Cisco: Cisco Visual Networking Index: Forecast and Methodology. http://www.cisco.com/en/US/solutions/collateral/ns341/ns525/ns537/ns705/ns827/white_paper_c11-481360.pdf (2012–2017)
3. SOCAP International, Now Arriving: Big Data in the Hospitality, Travel, and Tourism Sector. http://www.socap.org (2013)

4. LHC Guide, English Version: A Collection of Facts and Figures about the Large Hadron Collider (LHC) in the form of Questions and Answers, CERN-Brochure-2009-003-Eng. CERN. https://cds.cern.ch/record/1165534/files/CERN-Brochure-2009-003-Eng.pdf (2013)
5. Data, Data Everywhere, The Economist. http://www.emc.com/collateral/analyst-reports/ar-the-economist-data-data-everywhere.pdf (2012)
6. Delort, P.: OECD ICCP Technology Foresight Forum. http://www.oecd.org/sti/ieconomy/Session_3_Delort.pdf#page=6 (2012)
7. Webster, P.: Supercomputing the Climate: NASA's Big Data Mission. CSC World, Computer Sciences Corporation. http://www.csc.com/cscworld/publications/81769/81773-supercomputing_the_climate_nasa_s_big_data_mission (2013)
8. Davenport, T.H., Harris, J.G.: Competing on Analytics: The New Science of Winning. Harvard Business Review Press, Boston (2007)
9. IDC Digital Universe. Data, Bigger Digital Shadows, and Biggest Growth in the Far East. http://www.whizpr.be/upload/medialab/21/company/Media_Presentation_2012_DigiUniverseFINAL1.pdf (2012)
10. Davenport, T.H., Patil, D.J.: Data Scientist: The Sexiest Job of the 21st Century. Harv. Bus. Rev. **90**, 70–76 (2012)
11. Lund, S., Manyika, J., Nyquist, S., Mendonca, L.: Game changers: Five opportunities for US growth and renewal, McKinsey Global Institute Report. http://www.mckinsey.com//media/McKinsey/dotcom/Insights/Growth/US%20game%20changers/MGI_Game_changers_US_growth_and_renewal_Full_report.ashx (2013)
12. Laney, D.: 3D Data Management: Controlling Data Volume, Velocity and Variety, Gartner. http://blogs.gartner.com/doug-laney/files/2012/01/ad949-3D-Data-Management-Controlling-Data-Volume-Velocity-and-Variety.pdf (2001). Accessed 6 Feb 2001
13. Snijders, C., Matzat, U.: Big data: big gaps of knowledge in the field of internet. Int. J. Int. Sci. (2012). http://www.ijis.net/ijis7_1/ijis7_1_editorial.html
14. Laney, D.: The Importance of 'Big Data': A Definition, Gartner. http://www.gartner.com/DisplayDocument?id=2057415&ref=clientFriendlyUrl (2012)
15. What is Big Data? Villanova University. http://www.villanovau.com/university-online-programs/what-is-big-data/
16. Oracle, An Oracle White Paper: Information Management and Big Data. http://www.oracle.com/technetwork/topics/entarch/articles/info-mgmt-big-data-ref-arch-1902853.pdf (2013)
17. IBM. IBM Commercial The Road: Intelligent Data Management and Analysis. http://www.dailymotion.com/video/xdaoae_ibm-commercial-the-road-intelligent_tech#from=embediframe
18. Chanchal, Y., et al.: Algorithm and approaches to handle large data-a survey. Int. J. Comput. Sci. Netw. **2**(3), 37–41 (2013)
19. ELGAMA-ELEKTRONIKA, "GAMA 100 for residential and commercial metering" brochure. http://www.elgama.eu/templates/files/tiny_mce/Broshures/GAMA_100_EN_v4.4.pdf
20. ENERGA Operator, A.Olszewski: AMI Project Summary. http://www.meter-on.eu/file/2013/11/Energa.pdf (2013). Accessed 24–28 June 2013
21. Skoczkowski, T.: Smart metering services roadmap for Poland. Summary in English, Polish National Energy Conversation (2011)
22. Skoczkowski, T., Loth-Babut, K.: Work Package 3. Roadmap for innovative smart metering services in Poland. Project Progress Meeting, Bucharest (2012). Accessed 23–24 May 2012
23. http://www.pgedystrybucja.pl/spolka/aktualnosci/pilotazinteligentnego-systemupomiarowego-w-pge-dystrybucja. Accessed 03 Dec 2015
24. Rozwalka, T.: Budowa Infrastruktury Inteligentnego Pomiaru W PGE Dystrybucja SA, XX Forum Teleinformatyki. http://forumti.pl/20Forum/prezentacje/Tomasz_Rozwalka.pdfs. 10-11 (2014). Accessed 25–26 Sept 2014
25. ELGAMA-ELEKTRONIKA. Smart Metering Solution. http://www.elgama.eu/templates/files/tiny_mce/Broshures/AMI_EN_v2.pdf (2015). Accessed 27 Oct 2015
26. Polityka Energetyczna Polski Do 2030 Roku. https://www.mg.gov.pl/NR/rdonlyres/8C3B84AA-C9EF-471A-9B33-A66E2C700CC0/37509/PEPwer3210092007.pdf

27. Accenture. Realizing the Full Potential of Smart Metering. Accenture's Digitally Enabled Grid program. http://www.accenture.com/SiteCollectionDocuments/PDF/Accenture-Smart-Metering-Report-Digitally-Enabled-Grid.pdf
28. Nederland Netbeheer. Dutch Smart Meter Requirements, vol. 4. Netbeheer Nederland, Technical Report (2011) http://files.domoticaforum.eu/uploads/Smartmetering/DSMR%20v4.0%20final%20P1.pdf (2015). Accessed 17 Nov 2015
29. EEO Country Fact Sheet, Poland. http://ev-observatory.eu/module/factsheetpdf.php?rid=223&c=1

Intelligent System of Limited Resource Allocation for Large-Scale Agent Systems

Jakub Weclawski and Stanislaw Jankowski

Abstract This paper describes an intelligent decision support system for semi-autonomous agents. The solution is developed for a large-scale network of double-interface mobile routers. The network is a part of advanced metering infrastructure deployed in the north of Poland. Limited capacity of cellular networks makes the system stability and performance dependent on smart management of radio resource utilization. A model of system phenomena was based on observations of a real system consisting of more than 10,000 devices. A network simulator was developed as a tool for solution testing outside of the real system. Computational Intelligence methods were found suitable for a large system scale. The paper introduces an iterative method based on cellular neural network model for determining the optimal resource allocation for the system.

Keywords Intelligent decision support systems · Computational intelligence · Cellular neural network · Advanced metering infrastructure · Smart grid

1 Introduction

According to the Directives of the European Parliament and of the Council published in 2009 [3, 4], Member States of the European Union are required to ensure the implementation of intelligent metering systems to help consumers in active use of the market supply of electricity and gas. Systems of remote access to meters are organized in various ways, but their important component are transmission solutions

J. Weclawski (✉)
Institute of Electronic Systems, Warsaw University of Technology, Warsaw, Poland
e-mail: j.weclawski@stud.elka.pw.edu.pl

J. Weclawski
MindMade Ltd, Warsaw, Poland

S. Jankowski
Institute of Electronic Systems, Warsaw University of Technology,
Warsaw, Poland
e-mail: sjankows@elektron.elka.pw.edu.pl

D. Ryżko et al. (eds.), *Machine Intelligence and Big Data in Industry*,
Studies in Big Data 19, DOI 10.1007/978-3-319-30315-4_17

that allow for remote access from the acquisition server to the point where data is available [10, 12]. This task is typically implemented in cellular network technology. Requirements for improving the reliability of the readings results in the fact that transmission devices are equipped in additional communication interfaces. Their task is to ensure the communication in case of a significant deterioration of the operating parameters of the active interface. In that case, the device should switch to the other interface, which serves as backup. In large-scale deployments involving tens of thousands of devices, the configuration of leading technology for each of the system devices should be selected very carefully. Otherwise, there is a risk of overloading of network transmitters due to the limited capacity of cellular networks. That leads to unstable and sub-optimal operation of the system. It should also be noticed that the system operates in a dynamic environment. Signal parameters depend on the transmission traffic timing, the number of active subscribers and weather conditions. Each of these factors can cause device switching, especially when transmitters are permanently overloaded. To make the right choice of leading technology, it is not enough to observe signal parameters locally. It may happen that in order to ensure a high level of system availability, devices with good signal conditions should depart resources to their neighbors in less favorable situation. To do so, the device needs information about neighboring devices and the conditions under which they operate. Acquiring this knowledge by each of the devices would require continuous and frequent message exchange. The scale of this communication would result in an additional, unacceptably high traffic. An alternative to this model is a central system, which collects information from devices periodically and determines the optimal solution.

2 Central Configuration System

The central system is a server that collects an information from devices about available resources and their quality. In addition, it can determine which devices are adjacent to each other (neighborhood defined as sharing the same resources). Observing the devices behavior allows for extracting nodes which present low quality of service and evaluating the performance of the entire system.

2.1 Signal Conditions

Management of radio technology allocation requires mapping a variety of conditions that determine the availability and quality of a radio resource [7–9]. The transmitter signal is available only in a certain area defined by the cell range. Signal strength decreases along with increasing distance from the transmitter. At the same time, the signal strength itself is an insufficient criterion, because only after taking into account interference level it can be determined whether the connection is stable and ensures the appropriate bit rate. Interference level is proportional to the number of active users and traffic level. By periodical device polling, the central server obtains information about transmitters seen by devices and signal conditions offered by them. After gathering this information it is possible to map the system topography.

2.2 Device Control Mechanism

Management of resource allocation requires a device control mechanism. The one we propose here operates according to the following rules:

- if only signal conditions are good enough, use the technology set as the preferred one,
- if not, select the second one and wait until signal improves. Once it has improved, switch to the technology set as the preferred one.

Signal conditions are considered insufficient when signal strength or interference level falls below a determined threshold. When that happens, the device switches. It returns to priority technology if both signal parameters increase above thresholds. It should be noticed that the central system cannot withdraw the device's autonomy of reacting to sudden signal deterioration. Otherwise, the system would not be able to react on such sudden changes due to not having a communication path to forward its decision to the device.

2.3 System Configuration

The previous paragraph describes a mechanism for controlling the behavior of individual device by setting a leading technology. The priority shall be established by the central system and assigned to the device. It is only established once for passive systems, but in the case of active systems the priority setting constantly adapts to changing external conditions. While priority determines the settings for one device, configuration is defined as the priority setting for all of devices.

2.4 Global Measures of System Operation Quality

The described system works in a dynamically changing environment. It results in a continuous device switching and disconnecting, often characterized by oscillation. Switching is of course a natural mechanism of the system; however, its high level indicates system instability. The instability may be an effect of resource limitation, but also of improper resource allocation. In order to evaluate system behavior we propose two quantities. The first one determines the level of switches and the second one— the device disconnection level. Both of these values change dynamically over time, therefore must be seen in the long term perspective. Their average levels indicate the quality of the system operation. Two different configurations can be compared by observing the system behavior represented by these two measures.

In the model with discrete time, the average value of switch level is given by

$$\bar{\delta} = \frac{1}{N}\left(\sum_{n=1}^{N} \delta_{12}(n) + \sum_{n=1}^{N} \delta_{21}(n)\right) \tag{1}$$

where

N—number of epochs considered,
$\delta_{12}(n)$—number of switches from technology 1 to technology 2 at epoch n,
$\delta_{21}(n)$—number of switches from technology 2 to technology 1 at epoch n.

Analogously the average value of disconnection level is given by

$$\bar{\nu} = \frac{1}{N}\sum_{n=1}^{N} \nu(n) \tag{2}$$

where

$\nu(n)$—number of disconnected devices at epoch n.

2.5 Summary

The role of the central system is to determine the optimal configuration. This configuration should meet the following conditions:

- to ensure that all devices are connected,
- to provide the best operating conditions for all of the devices,
- to minimize the number of switches.

3 Simulator of Large-Scale System of Double-Interface Routers

In order to develop an effective resource allocation method, a network simulator was proposed as a tool of system exploring and solution testing. Since in a real system devices switch, lose connection and transfer data, it was crucial to model the system evolution over time.

3.1 Device, Transmitter and Area Models

Area model—2-dimensional grid with devices and transmitters position determined. The area is a rectangular matrix in which each cell can contain a device and/or a transmitter. For each grid cell all visible transmitters are identified.
Device model—the main goal of the device is to stay connected. It has two radio interfaces at disposal. At any given time, only one is used to transmit data, while the other acts as a backup interface. Device status determines which interface is currently the active one. If the device fails to attach a session on any of the interfaces, then the state is set as "disconnected".
Transmitter model—transmitter provides a signal of a given technology. The range determines the coverage area of the transmitter. In the simplest model, it is a wheel-shaped cover centered on the transmitter. Its size is dependent on the transmission power of the transmitter. It can be assumed that for a given technology the range is constant and its initial value is dependent on the frequency band of the corresponding technology. Range coverage can vary according to traffic load.
Grid cells R_n in the range cover of transmitter t_n are given by

$$R_n = \{m_{ij} \in M : dist(m_{ij}, t_n) \leq r_n\} \tag{3}$$

where

m_{ij}—grid cell at position $< i, j >$,
M—a set of grid cells,
t_n—transmitter t_n,
r_n—transmitter t_n range.

3.2 Transmitter Overload Model

Connected devices are divided into those that use a particular technology as the leading one, and those for which it is a backup technology. The second type of connection does not result in an increased transmitter load. For each transmitter two limits are defined. A warning limit l_w indicates the maximum device quantity for which a particular technology is the leading one with the transmitter still operating stably. The second one—a critical limit l_c—determines the maximum quantity of devices that can use the transmitter as the leading technology. However, their activity is already unstable and the device may be disconnected at any time. As a result, the number of active devices connected to the transmitter can vary from 0 to l_c, wherein when the value exceeds l_w, $l_w < l_c$, there is a risk of sudden disconnection. When the number of devices tends to l_c, the risk of disconnection increases. When the number of active devices equals to l_c, new devices are no longer being connected to the transmitter. Probability of disconnection for the proposed model is then given by (Fig. 1)

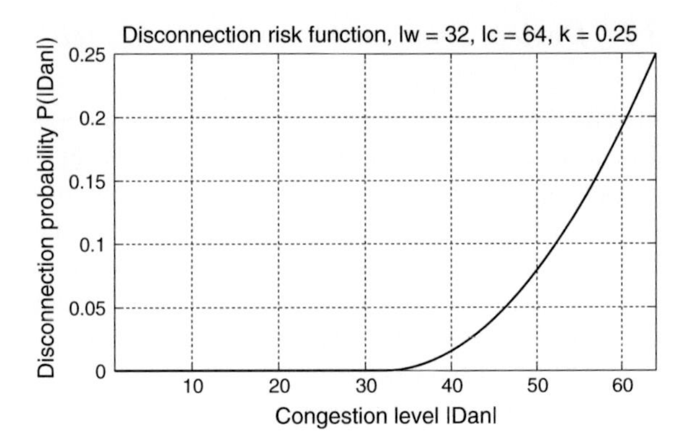

Fig. 1 Probability of device disconnection in function of transmitter congestion level $|D_{an}|$

$$P(|D_{an}|) = \begin{cases} 0 & \text{for } |D_{an}| \le l_w \\ k|D_{an}| \left(\frac{|D_{an}|-l_w}{l_c-l_w}\right)^2 & \text{for } l_w < |D_{an}| \le l_c \end{cases} \tag{4}$$

where

l_w—warning limit,

l_c—critical limit,

D_{an}—a set of devices, for which t_n is an active technology transmitter,

$|D_{an}|$—a number of elements of a set D_{an}, identical with a congestion level of transmitter t_n,

k—coefficient of the maximum disconnections level, $k = (0, 1\rangle$.

3.3 *Interface Model*

The interface binary state is defined as:

- connected—the interface is connected,
- disconnected—the interface is disconnected because of the lack of radio resources. The lack of resources can mean one of two situations (i) the lack of coverage for a given technology (ii) weak signal conditions indicating the overload of nearby transmitters.

s_n^1—technology 1 interface state for device d_n, $s_n^1 = \{0, 1\}$ (0: disconnected, 1: connected), is given by

$$s_n^1 = \begin{cases} 1 & \text{if } T_{ij}^1 \neq \emptyset \wedge \exists t_n \in T_{ij}^1 : |D_{an}| < l_c \\ 0 & \text{if } T_{ij}^1 = \emptyset \vee T_{ij}^1 \neq \emptyset \wedge \forall t_n \in T_{ij}^1 : |D_{an}| = l_c \end{cases} \tag{5}$$

s_n^2—technology 2 interface state for device d_n, $s_n^2 = \{0, 1\}$ (0: disconnected, 1: connected), is given by

$$s_n^2 = \begin{cases} 1 & \text{if } T_{ij}^2 \neq \emptyset \wedge \exists t_n \in T_{ij}^2 : |D_{an}| < l_c \\ 0 & \text{if } T_{ij}^2 = \emptyset \vee T_{ij}^2 \neq \emptyset \wedge \forall t_n \in T_{ij}^2 : |D_{an}| = l_c \end{cases} \quad (6)$$

where

T_{ij}^1—a set of transmitters of technology 1. available at position $< i, j >$,

T_{ij}^2—a set of transmitters of technology 2. available at position $< i, j >$,

D_{an}—a set of devices, for which t_n is an active technology transmitter,

$|D_{an}|$—a number of elements of a set D_{an}, identical with a congestion level of transmitter t_n.

Interface role:

- leading interface—the interface is used for data transmission. To fulfill this function, the interface must be able to "connect". At any given time, only one of the interfaces can be the leading one, while the other one acts as a backup interface.
- back-up interface—used when the current leading interface loses connection due to deterioration of signal conditions and is no longer able to fulfill its function. In such case the backup interface takes the role of the leading one. When the leading interface works correctly, the backup one remains dormant.
- priority interface—prioritization decides which of the interfaces should be used as the leading one. If the interface priority can not fulfill its function, the backup interface takes over this role until signal conditions for priority interface improve.

The device status s_n, $s_n = \{0, 1, 2\}$ (0: disconnected, 1: technology 1, 2: technology 2), is given by

$$s_n = \begin{cases} p_n & \text{if } s_n^{(p_n)} = 1 \\ \neg p_n & \text{if } s_n^{(p_n)} = 0 \wedge s_n^{(\neg p_n)} = 1 \\ 0 & \text{if } s_n^{(p_n)} = 0 \wedge s_n^{(\neg p_n)} = 0 \end{cases} \quad (7)$$

where

p_n—priority technology for device d_n, $p_n = 1, 2$ (1: technology 1, 2: technology 2). The logic for priority p_n is defined as: for $p_n = 1$ $\neg p_n = 2$, for $p_n = 2$ $\neg p_n = 1$.

3.4 *Simulator*

The main task of the simulator is to emulate the system evolution, which includes devices' connecting, switching and losing connection due to cellular network transmitter overloading. This was achieved by repeating two main mechanisms of the simulator. The first was associated with initialization and renewal of the device connection and the second was associated with the transmitter's response to overloading. The operation of the simulator can be described as a pseudo-code shown below.

Pseudo-code of the simulator

```
initialization
loop(number of epochs)
{
     disconnections phase
     renew connections phase
}
```

3.5 Results of System Simulation

The results of an exemplary simulation are presented below. Evolution of a system consisting of 2400 devices was observed. The simulation was carried out for 50 epochs. Figure 2 shows the state of the system at the beginning and in the end of the simulation. Particular cells of the map represent devices that can be connected to the first or the second technology (blue and orange), or disconnected when run out of resources (black).

Figure 3a presents a network coverage map. Blue and orange indicate individual technologies. Black marks cells not covered by any transmitter. White indicates that both technologies are available in a given cell. Figure 3b shows system evolution in number of switches and level of disconnected devices. According to what was suggested earlier, these levels can vary significantly in every epoch.

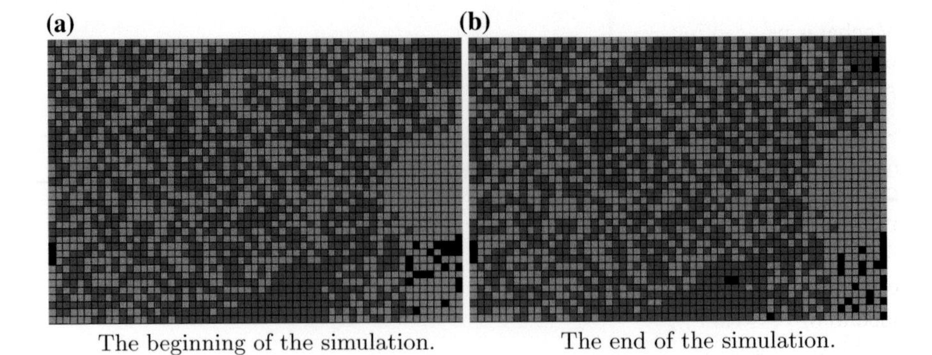

(a)
(b)
The beginning of the simulation.
The end of the simulation.

Fig. 2 The system state at the beginning and in the end of the simulation. *Blue* ■ and *orange* ■ indicate the allocation of each technology. The *black* ■ marks disconnected devices

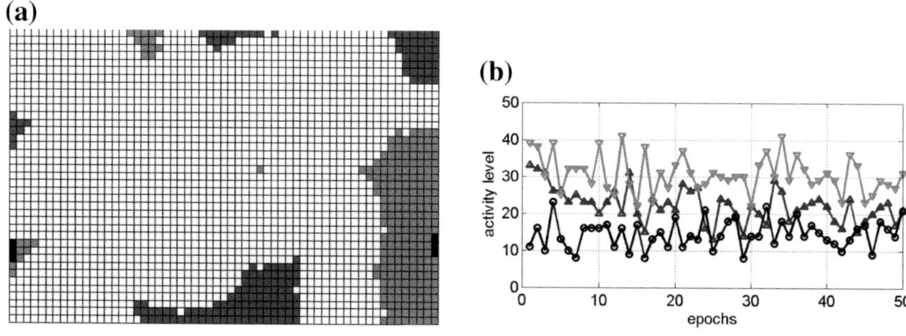

Network coverage map. Blue ■ and orange ■ indicate individual technologies. Black ■ marks cells not covered by any transmitter. White □ indicates that in a given cell both technologies are available.

Switch and disconnection level evolution. Blue △ - switch to technology 1, orange ▽ - switch to technology 2, black ○ - disconnection level.

Fig. 3 Network coverage map (**a**). Switch and disconnection level evolution (**b**)

4 System of Resource Allocation

In chap. 3 the simulator was presented. It allows the system evolution observation for a given system configuration. This chapter presents a method of determining the optimal configuration.

4.1 Optimal Configuration Conditions

Mechanisms used in the simulator are based on phenomena occurring in real networks. The quality of a network signal depends on the signal strength and the interference level. The signal strength can be seen as a feature of the area in which the device is located. It is sensitive to weather conditions and increased traffic. In a simplified model we consider here, it cannot be controlled and improved by configuration management. We also assume that if a device is within a transmitter's range, it means that the signal strength is sufficient for stable operation. On the other hand, the interference level is dependent on the number of active devices and on traffic generated by them. As opposed to the signal strength, it can be controlled by appropriate technology distribution. In the model shown in the previous section, the transmitter load level is associated with the number of connected devices. According to Eq. 4, switches in the system are proportional to the number of congested transmitters and their congestion level $|D_{an}|$:

$$\bar{\delta} \simeq \sum_{t_n \in T : |D_{an}| > l_w} k|D_{an}| \left(\frac{|D_{an}| - l_w}{l_c - l_w} \right)^2 \quad (8)$$

Disconnections are proportional to the number of devices which have no access to technology because they are beyond any transmitter's range or all available transmitters are overloaded $|D_{an}| = l_c$:

$$\bar{\nu} \simeq |d_{ij} \in D : T_{ij} = \emptyset| + |d_{ij} \in D : T_{ij} \neq \emptyset \wedge t_n \in T_{ij} : |D_{an}| = l_c| \quad (9)$$

where

d_{ij}—device at position $< i, j >$

If we assume that the transmission traffic is uniformly distributed between network transmitters, the optimal configuration is the one that ensures equal transmitter load. According to the adopted model, it will be equivalent to the uniform distribution of devices between transmitters. According to Eq. 8 equal transmitter load minimizes the number of switches. At the same time, it reduces the second term of Eq. 9 by avoiding an excessive overloading of network transmitters (the first term will remain constant regardless of the configuration). Averaging the transmitters load level also averages the interference level in the system. As a result, it provides the best signal conditions for devices in the scale of a whole population. In the end, the designated optimality condition realizes all requirements postulated in Sect. 2.5 for centralized configuration system. The question remains as to provide such a distribution in case of complex and heterogeneous system topography.

4.2 *Cellular Neural Networks for Optimization*

In order to solve the problem formulated, a model based on cellular neural network [1, 2, 5] was proposed. This model allows for mapping a system's dependency resulting from both the transmitters' and devices' topography and the interactions between them. In the model, we introduced a two-layer node grid. The first layer reflects the spatial arrangement of devices, and the second transmitter and radio signal distributions. Both layers are composed into a two-dimensional map, which allows for referring spatial relationships of these two layers. Interactions between devices resulting from overloading a shared resource propagate through transmitters. At the same time, limited transmitter ranges result in the fact that these interactions are local—limited to devices sharing the same resource (transmitter). Node status is identical to the identifier of the transmitter providing signal for the active interface. In the final solution, the priority will be determined as a technology associated to a selected transmitter, but for the need of solution finding, the state is identical to the transmitter identifier. Dynamics of node state changes is given by

$$s_n = \arg \min_{t_n \in T_{ij}} (|D_{an}|) \quad (10)$$

indicating that the device will change its state (i.e. select another transmitter) if there is another transmitter available at device position offering better signal, which in the proposed model is identical to a lower load level. The allocation of preferred technologies (selection of the transmitter) can also be understood as a type of clustering task [11], in which devices are divided into clusters with cellular transmitters acting as the attraction points. Such clusters are grouped spatially (including devices in the range of a single transmitter), but at the same time they are locally distributed between transmitters to avoid single resource overloading.

4.3 Method of Determining the Optimal Configuration

We propose an iterative method of system adaptation [6] which evolves from the initial, unstable state to the final, optimal configuration. It is accomplished by switching devices to transmitters with the lowest load level. If in every iteration a part of the population selects transmitters with minimum load level, then in the end we will get a uniform distribution of the devices. If at this point we attribute priorities to devices according to the technology selected in the process of transmitter optimization, we can expect that this configuration will ensure the optimal performance of the system.

Pseudo-code of proposed algorithm

```
initial configuration
loop(number of epochs)
{
```

$$s_{ij} = \arg_{t_n \in T_{ij}} \min(|D_{an}|)$$

```
}
```

where

s_{ij}—state of the device on position $< i, j >$, identical with transmitter ID that is used by device active interface,

T_{ij}—the set of the transmitters available at position $< i, j >$.

4.4 Results of Configuration Optimization

Below we present results of an exemplary run of the algorithm. The configuration was determined for 2400 devices and two available technologies. Figure 4 shows the initial and final system configuration. Blue and orange indicate the allocation of individual technologies. Devices marked with black are beyond the reach of any of the technologies. Figure 5 shows the evolution of the system and the process of reaching stability. This process can be tracked by observing the standard deviation of the transmitters load level. The lower it is, the more uniform the device distribution between transmitters (Fig. 5a). Along with equalizing the transmitters load level, the number of switches also decreases (Fig. 5b).

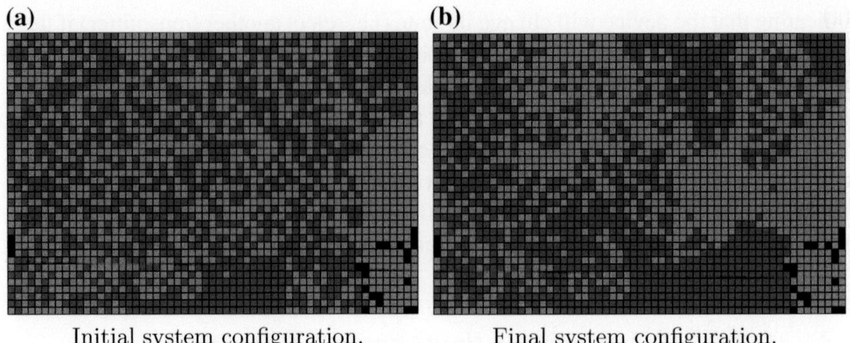

Initial system configuration.

Final system configuration.

Fig. 4 Initial (**a**) and final (**b**) system configuration. *Blue* ■ and *orange* ■ indicate the allocation of each technology. *Black* ■ indicates devices beyond the reach of any of the technology

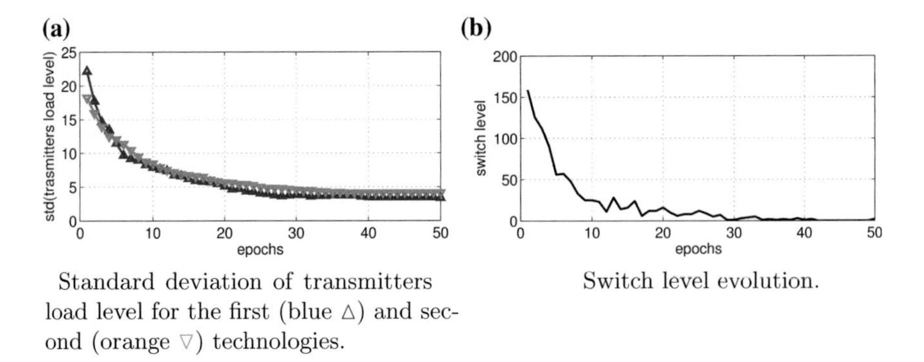

Standard deviation of transmitters load level for the first (blue △) and second (orange ▽) technologies.

Switch level evolution.

Fig. 5 Standard deviation of transmitters load level evolution (**a**). Switch level evolution (**b**)

5 Evaluation of Different System Configurations

The simulator was used to evaluate the method proposed. It was done by comparing the system behavior for two different system configurations: the random one and the one determined by the method described. In the case of a random configuration, the technology distribution should reflect the ratio between the number of transmitters available for these two technologies. As the amount of transmitters was set to the same value, the devices were divided evenly—half prefers the first technology, half the second (devices were selected randomly). If leading technology was set as the one that was not available in the device area, the device would operate stably, unlike it might be expected. The backup interface would be set as leading one and the device would not try to switch. In this respect, a random configuration is not pointless. Of course, in areas where both technologies are available, the priority distribution would be very important and it would decide about the optimality of resource utilization. The two configurations considered are shown in Fig. 6. In the second chapter we

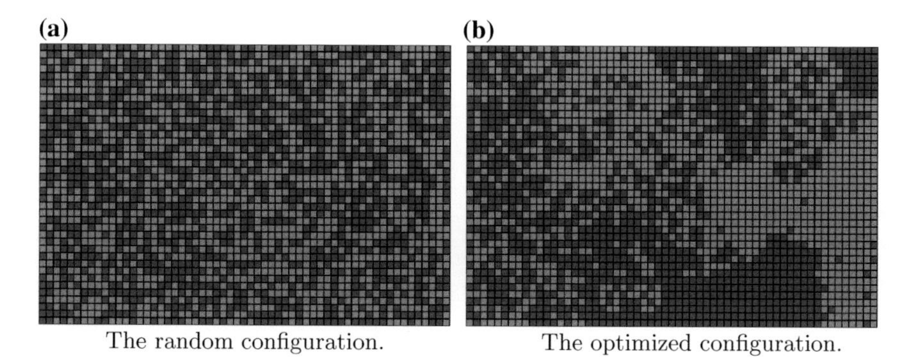

Fig. 6 Two different configurations, the random one on the left, the optimized one on the right. *Blue* ■ and *orange* ■ indicate the allocation of each technology

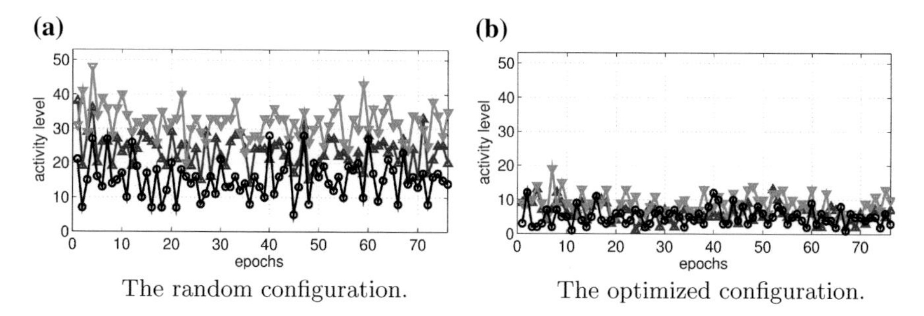

Fig. 7 System evolution comparison for two proposed configurations. On the left the random one, on the right the optimized one. *Blue* △—switch to technology 1, *orange* ▽—switch to technology 2, *black* ○—disconnection level

Table 1 Switch ($\bar{\delta}_{12}, \bar{\delta}_{21}$) and disconnection ($\bar{\nu}$) level for two different configurations (i) the random one (ii) optimized by the proposed method

Configuration		$\bar{\delta}_{12}$	$\bar{\delta}_{21}$	$\bar{\nu}$
Random	Mean level	21.6	29.8	15.1
	Standard deviation	6.6	9.4	5.2
Optimized	Mean level	4.8	5.6	4.6
	Standard deviation	2.9	3.0	2.5

defined measures that allow to assess the quality of system behavior. We divided it into two quantities: the average level of switches and the average level of disconnected devices. Figure 7 shows the evolution of the system in the period of 75 epochs of the simulation. Table 1 shows the average values of switch and disconnection level during the entire simulation. The presented results show that the method allowed for reducing the switch and disconnection level by factor of five and three respectively.

Analogous results were obtained for all the simulations carried out. Because of the lack of space we present only one example, which is however a representative simulation.

6 Conclusions

The paper presents a method of determining the optimal system configuration related to the problem of radio resource management for a system of double-interface mobile devices. This system can be viewed as a large-scale network of coupled binary switches. A model of mutually interacting devices and cellular network transmitters was proposed to map a system dynamics. We present a simulator of a large-scale network of mobile devices, which is a tool to study and evaluate the performance of the system for its various configurations. We propose a model based on a cellular neural network that was used to determine the optimal system configuration. It was proposed as the perfectly suited tool for mapping relationships resulting from the system topography, mutual interactions and system dynamics. It was shown that efforts should be undertaken to load cellular transmitters uniformly in order to avoid a situation in which the network operates in a state of permanent overload. Otherwise the system would operate unstably. Moreover, this may cause disconnection of devices that despite having access to several radio technologies are not able to attach a connection. The method proposed allows for a significant improvement in radio resource utilization quality. As it was pointed in table 1, the results were improved by reducing the numbers of switches and disconnections from 22 to 5, from 30 to 6 and from 15 to 5 respectively. At the same time, by avoiding excessive transmitter load, we provide a margin that ensures network security in the situation of possible increased load level. The optimization was performed assuming uniform transmission traffic and without modeling the phenomenon known as "cell breathing", which names cell coverage reduction resulting from increased traffic. Although at this moment we do not include these conditions, the proposed model is perfectly suited for their inclusion. Moreover, the conclusions based on condition of uniform transmitter load level are also valid for this new, more advanced model. In the future work, the expansion of the optimum criterion is needed. We plan to expand it as to provide not only uniform transmitter load level but also a maximum signal strength. If we assure that devices do not work on the transmitter range boundary we make the system immune to cell coverage reduction consequences. In the next step, we also want to propose a cooperative model, in which devices with good signal conditions depart resources to their neighbors in less favorable situation. Although in this work we analyze only a system which is a part of advanced metering infrastructure, the research conclusions are also valid for other cellular systems, in which various technologies are available. The idea of smart radio resource management by technology configuration we also find useful for Internet of Things solutions.

The authors gratefully acknowledge to MindMade Ltd for providing their infrastructure for the purposes of this research and to Jan Jakub Szczyrek for inspirational discussions that were very helpful in appropriate problem formulation.

References

1. Chua, L.O., Yang, L.: Cellular neural networks: theory. IEEE Trans. Circuits Syst. **35**(10), 1257–1272 (1988)
2. Chua, L.O., Yang, L.: Cellular neural networks: applications. IEEE Trans. Circuits Syst. **35**(10), 1273–1290 (1988)
3. Directive 2009/72/EC of the European Parliament and of the Council of 13 July 2009 concerning common rules for the internal market in electricity, http://eur-lex.europa.eu
4. Directive 2009/73/EC of the European Parliament and of the Council of 13 July 2009 concerning common rules for the internal market in natural gas, http://eur-lex.europa.eu
5. Fantacci, R., Forti, M., Marini, M., Pancani, L.: Cellular neural network approach to a class of communication problems. IEEE Trans. Circuits Syst. **46**(12), 1457–1467 (1999)
6. Herault, L.: Neural networks without training for optimization, In: Dreyfus, G. (ed.) Neural Networks, pp. 443–486 Springer, New York (2005)
7. Kim, Kiseon, Koo, I.S.: CDMA Systems Capacity Engineering. Norwood, USA: Artech House (2004)
8. Kolakowski, J., Cichocki, J.: UMTS—Third Generation Mobile Cellular System. WKL, Warsaw (2009). (in Polish)
9. Lempiainen, J., Manninen, M. (eds.): UMTS Radio Network Planning. Optimization and QoS Management. For Practical Engineering Tasks, Springer, USA (2003)
10. Parikh, P., Kanabar, M., Sidhu, T.: Opportunities and challenges of wireless communication technologies for smart grid applications. In: Proceedings of the 2010 IEEE Power and Energy Society General Meeting, pp. 1–7
11. Xu, R., Wunsch II, D.: Survey of clustering algorithms. IEEE Trans. Neural Netw. **16**(3) (2005)
12. Yan, Y., Qian, Y., Sharif, H., Tipper, D.: A survey on smart grid communication infrastructures: motivations, requirements and challenges. IEEE Commun. Surveys Tuts. **15**(1), 5–20 (2013)

[illegible]

References

[illegible]

Searching for Logical Patterns in Multi-sensor Data from the Industrial Internet

Mohit Yadav, Ehtesham Hassan, Gautam Shroff, Puneet Agarwal and Ashwin Srinivasan

Abstract Engineers analysing large volumes of multi-sensor data from vehicles, engines etc. often seek to search for events such as "hard-stops", "lane passing" or "engine overload". Apart from such visual analysis for engineering purposes, manufactures also need to count occurrences of such events via on-board monitoring sensors that ideally rely on classifiers; searching for patterns in available data is also useful for preparing training sets in this context. In this paper, we propose a method for searching for multi-sensor patterns in large volumes of sensor data using qualitative symbols (QSIM (Say, Functions representable in pure QSIM, 251–255, 1996, [1])) such as "steady", "increasing", "decreasing". Patterns can include symbol-sequences for multiple sensors, as well as approximate duration, level or slope values. Logical symbols are extracted from multi-sensor time-series and registered in a trie-based index structure. We demonstrate the effectiveness of our retrieval and ranking technique on real-life vehicular sensor data in the visual analytics as well as classifier training and detection scenarios.

Keywords Pattern search · Time-series · Multi-sensor data · Multivariate time series

---

M. Yadav (✉) · E. Hassan · G. Shroff · P. Agarwal
TCS Research, New Delhi, India
e-mail: y.mohit@tcs.com

E. Hassan
e-mail: ehtesham.hassan@tcs.com

G. Shroff
e-mail: gautam.shroff@tcs.com

P. Agarwal
e-mail: puneet.a@tcs.com

A. Srinivasan
BITS, Goa, India
e-mail: ashwin.srinivasan@wolfson.oxon.org

D. Ryżko et al. (eds.), *Machine Intelligence and Big Data in Industry*,
Studies in Big Data 19, DOI 10.1007/978-3-319-30315-4_18

1 Introduction

A notable development in the field of vehicle sensor technology has been due to the increasing availability of large volumes of sensor data from field-test as well as production machines. Modern vehicles, and in general machines of all kinds (engines, factory equipment etc.) are being instrumented with a large number of sensors to capture detailed information about their behaviour for further analysis. Increasingly such sensors are also able to regularly transmit their data to the component (e.g. engine) manufacturer, OEM (e.g. aircraft/car manufacturer) or even operator (e.g. airline/trucking company): This phenomenon now being referred to as the *industrial internet* [2].

While engineering new machines of all kinds, quality engineers schedule trial runs in a controlled environment to simulate the life cycle of a machine in the field. In order to do this effectively, one must understand how similar models are used in the field by customers, especially in terms of the distribution of certain manoeuvres that stress various components, such as rapid acceleration, sudden braking, sharp turns, etc. In this paper we present a generic method of searching for such events from voluminous multivariate time-series data arising out of the industrial internet.

Events of interest are often characterised by a set of patterns in multiple sensors, occurring simultaneously. Further, the duration of every occurrence of an event is not constant, for example, vehicle drivers don't always apply sudden brakes for a fixed duration every-time. Not just this, the duration of the event is observed to be different in every sensor, involved in the event; for example, when applying sudden brake in a vehicle the pattern in brake sensor is of longer duration than that in speed sensor. The problem becomes even more complex because patterns need not occur in the same time-window for each sensor: for example, there may be a small time gap between start of a 'sudden brake' pattern in speed sensor versus the brake sensor. Missing values in sensor readings further complicate the matter. Thus, searching for multi-sensor patterns becomes challenging in practice.

Engineers often define the events they are interested in based on domain knowledge. However, it then requires significant manual effort to annotate events in voluminous sensor data. In practice, engineers detect such events by programmatically searching for hard-coded conditions being observed in the sensor values; for example, an event might be translated into a hard-coded rule seeking situations where "sensor-1 is higher than s_1 for duration t_1 and derivative of sensor-2 is more than d_2 for duration t_2". Because of the practical challenges mentioned earlier, such hard-coded rules are often ineffective [3], as well as difficult to implement. Alternative techniques for the detection of time-series events [3–5] use machine-learning techniques, but most of these methods work on time-series subsequences that have been extracted from the whole time-series already. Most of these methods do not critical aspects of the problem described above, such as variable duration, mis-alignment in time.

Our main contributions are as follows: (1) We submit that events of practical interest can often be described as the simultaneous occurrences of *logical patterns* across multiple sensors. We use logical symbols such as *increasing* '+', *steady* '0', or *decreasing* '−' in terms of slope, as this allows us to describe patterns in easy and succinct manner. (2) We propose a query language based on these logical symbols using which engineers define a multi-sensor event. (3) We also describe how time-series of each sensor can be converted into a sequence(often much smaller) of logical symbols. (4) We also describe a scheme to store and index each sensor time-series in terms of such logical patterns. (5) An algorithm for searching and ranking instances retrieved against query defined using proposed query language. Our method of searching for multivariate time-series events addresses a number of practical challenge including but not limited to, variable duration of events, shift in the time of occurrences of patterns across different sensors, and variability in the response time of different sensors to the same actual event.

In practice, we found our technique can be utilized in multiple scenarios: (A) Our technique provides a mechanism for exploratory analysis of voluminous multi-sensor data, wherein occurrences of logically defined events are retrieved for further visual analysis by engineers. However, depending on how a logical query is framed, it can result in spectrum of precision-recall behaviours: In the case of scenario A above, reasonable recall with high precision is sufficient for exploratory purposes. However, in case when one wants to discover *all* events, a machine-learning approach such as [3, 4] is desirable. Still, even to use such techniques, one first needs to provide a training set: (B) Our search technique can be used to filter a small set of candidate set of potential event occurrences which can then be human-tagged to create a training set for these machine-learning approaches. Further, when applying one of these machine-learning-based classifiers, we usually need to provide it with a set of candidate *sub-sequences*: (C) Using generously specified queries to increase recall, our search technique can be used as a pre-filter to extract sub-sequences followed by applying a machine-learning-based classifier such as [3, 4].

The remainder of the paper is organised as follows: We first place our contributions in the context of related work in Sect. 2. In Sect. 3 and then describe our overall approach along with the scenarios in which it has been found useful in practice. Next in Sect. 4, we describe our technique logical pattern search in detail, including the query language, pre-processing, storage, retrieval and ranking. We illustrate our experimental results in Sect. 5 and conclude in Sect. 6.

2 Related Work

In the last decade, a tremendous growth of interest has been noticed in research and applications of time-series data [6–8]. Various attempts have been made in the past to convert time-series into symbolic representation [6, 9–13]. Ding et al. in [8] compare such methods of symbolic representation and indexing time-series data. Rakthanmanon et al. in [13] present a method for searching very long time-series.

Such efforts have greatly enhanced the state-of-art in this area in research domain. Most of these approaches focus on finding approximate matches provided one such time-series pattern. These approaches aim to find matches for a given pattern, not all those instances that might correspond to one such event. It is important to note that text based logical queries are more powerful in describing events in comparison to query by example. For example, events such as 'primary master cylinder pressure rises approximately at the rate of 10 psi roughly for 3 s followed by steady behaviour at or greater than 500 psi' are hard to search using such approaches, which is not the case with our approach.

Although, researchers have given attention to address challenges in dealing with patterns appeared in multivariate time-series data such as, Gillian et al. in [14] use DTW for identification of multivariate musical gestures. Some researchers model this problem as that of unsupervised frequent pattern discovery in time-series [15–19]. Objective behind these approaches is to discover frequent temporal patterns that provide important information about the behaviour of underlying system. In our case, objective is just reverse, i.e., to find instances of important patterns given information about their behaviour.

Readers are referred to [3–5, 20–23] for supervised classification based approaches for event detection in time-series data. List of other papers [24, 25] that use a mix of supervised and unsupervised methods to discover temporal patterns which can discriminate events from rest part of time-series data. But all such methods require label information in order to learn the behaviour of temporal patterns corresponding to such events. Also, most of these approaches work on time-series segments of fixed-length, but are capable of identifying patterns that not necessarily occur at fixed distance from the starting point of the time-series segments. However, these methods also struggle in detecting events, as they have to consider multiple lengths of windows which makes this task computationally expensive.

3 Framework Overview and Context

Figure 1 depicts the overview of proposed framework, in which a pattern description is expected to be queried to find instances with in time-series data. The pattern description should be in the form of symbolic query defined using our query language. We search and rank all hits retrieved by our system in response to a query. To facilitate efficient execution of such queries, time-series are pre-processed into sequences of logical symbols, which are clustered and stored on disk ('database summary'), as well as maintained in memory in a trie structure that is used to retrieve only those elements from disk that are required to answer a query at run-time. In practice our system has the been found mainly useful in the following practical contexts:

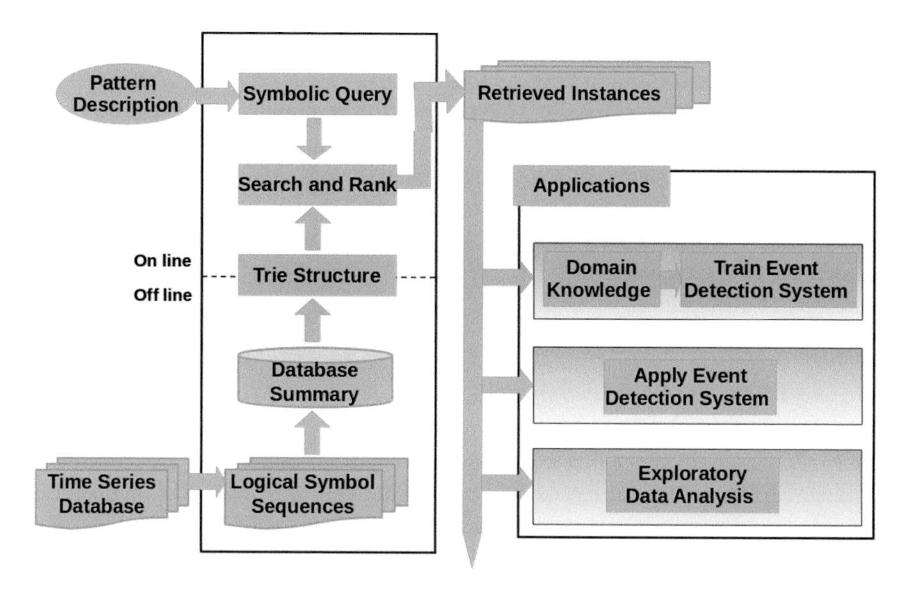

Fig. 1 Overview of the proposed framework

1. Annotating sufficient instances of *events* e.g. hard stops, layden/unlayden which can be further consumed in training a machine-learning system. Then a machine learning system can be utilized to detect all such events from the time-series data.
2. During test stage classifier can be applied only to sub-sequences retrieved against a pattern query. To ensure a higher recall, one can define the query in a generous manner. Which allows our technique to bypass the problem of selecting an optimal window size and also obviates the need to apply a classifier on every choice of such window or on every subsequence of a large time-series.
3. Exploratory data analysis: as a component of visual analytics as well as in several ways, e.g. in identifying rare/frequent patterns in time-series data. This information is useful in understanding behaviour of vehicles and in coming up with better designs in future.

4 Logical Pattern Search

Let us denote the multi-sensor time-series data obtained from multiple sensors by T. The time-series data T is recorded for many sessions of operation for the underlying mechanical system (e.g., run of a car, or a day of operation of a plant), over a period of months or years. We also refer to these sessions of operations as runs of underlying vehicles, or segments of the time-series T. Before processing this raw data T, we split data into multiple segments $\{T_1, T_2, ..., T_M\}$ which correspond to different runs of vehicles. We then convert every time-series segment T_i into a sequence of logical symbols $Z_i = \{Z_1^i, Z_2^i, ..., Z_N^i\}$ as explained in Sect. 4.1. In Sect. 4.2, we introduce a method for clustering and storage of logical patterns for every sensor. After that we explain trie-based index-structure, which is used to store these logical patterns. This trie based index-structure is used for search of multivariate events. Next, in Sect. 4.3, we explain proposed query language that can be used to describe multivariate events. In Sect. 4.4, we explain algorithm used to find matches of specified logical patterns across multiple sensors. Lastly, in Sect. 4.5, we present a method for ranking instances retrieved against a logical query.

4.1 *Logical Symbolic Encoding*

For every run T_i, firstly we compute slopes which is nothing but the difference between adjacent sensor values corresponding to every time-stamp in time-series. Next we take a sufficiently large number of samples of slopes and sort them. We then divide this sorted set into three groups with equal support followed by computing mean slope for every group. Based on the value of mean slopes, we represent these groups by symbols '+' '0' and '−', representing *increasing*(+), *steady*(0), and *decreasing*(−) patterns in time-series. We then compare the slope of every point in the multi-variate time-series T_i with mean slopes, and based on distances from them we assign that particular time-stamp as '+' '0' or '−'.

Algorithm 1: Merging criterion

```
1  while list Z_i has not reached end do
2      /% For every symbol Z^i_k %/
3      if length of Z^i_k ≥ duration threshold or Z^i_k is first/last symbol of series then
4          Shift to next symbol;
5          continue;
6      end
7      if symbol of Z^i_{k-1} is same as of Z^i_{k+1} then
8          Merge Z^i_{k-1}, Z^i_k and Z^i_{k+1} into single new triple with symbol of Z^i_{k-1};
9          Compute updated length and average slope for new triple ;
10         shift to next symbol;
11     else
12         if Z^i_k == '0' then
13             Merge Z^i_k to previous/after, whichever has slope nearer to slope Z^i_k;
14             Compute updated length and average slope for new triple ;
15             shift to next symbol;
16         else
17             Assign Z^i_k ← '0' and merge with '0' only if average slope of Z^i_k lies between center of symbol of
               Z^i_k and center of '0';
18             Compute updated length and average slope for new triple ;
19             shift to next symbol;
20         end
21     end
22 end
```

All continuous occurrences of a symbol are replaced by a pair of terms including the symbol and its length. In these pair of terms, we also include the average slope of various points that were removed to create this pair. This triple, is represented as Z^i_k, i.e., kth symbol from ith segment of time-series and it includes three terms: the symbol, its length and the average slope of points represented by Z^i_k. As a result, the time-series is represented by a sequence of logical symbols $Z_i = \{Z^i_1, Z^i_2, ..., Z^i_N\}$. We refine this logical sequence by merging triples that are of lesser duration than a threshold as explained in Algorithm 1. The intuition behind this additional refinement is to weed out noise. We merge them into symbols present at adjacent centroids if the symbol is not too far from it.

4.2 Symbolic Storage on Disk and in Memory

A *logical pattern* is a sequence of symbols '+', '0' and '−' without any pair of consecutive similar symbols. We extract all logical patterns upto a fixed length (usually taken as 5). After extracting logical patterns from sequences, we reject uninteresting patterns, which have *magnitude deviation* smaller than a prior threshold. The magnitude deviation of a logical pattern is measured as the difference between maximum and minimum z-normalized values of the time-series with in duration of that pattern. Since different sensors have different range of values, we perform Z-normalization to avoid multiple such thresholds.

After removing uninteresting patterns, we collect all instances that belong to one logical pattern, e.g., $\{-, 0, +\}$. Every pattern is represented by fixed length vector. This vector includes lengths (duration) and average slopes or magnitude(in case of '0') of every symbol in that pattern. These feature vectors are normalized to get rid of bias that comes because of high variation in their range. We then cluster these vectors using Birch clustering [26]. The distance measure used for clustering is similar to Euclidean distance, with a difference that if a pattern starts or ends with '0' symbol, its length is ignored. Intuition behind ignoring duration of '0' is motivated by the fact that a vehicle remains stationary for longer intervals of time after/before various events which is not relevant in many practical scenarios e.g. "It does not matter how long you have been driving before applying a hard stop".

We maintain a folder for every sensor and a file in that folder for every logical pattern. All clusters of a pattern are stored in the same file in sequential order. Each cluster description contains the vectors of the logical patterns, sorted according to index of time-series segment contained in it. For every logical pattern, we store durations and values of every symbol present and index with respect to that time-series segment. Even in time-series segment logical patterns are stored based on their index with in that segment. There is a exception for logical patterns which start with '0', for them we keep index of the next symbol present in logical pattern. During search, if we do not to match the index of '0' symbol if it has appeared at the first place rather we match index of symbol next to it.

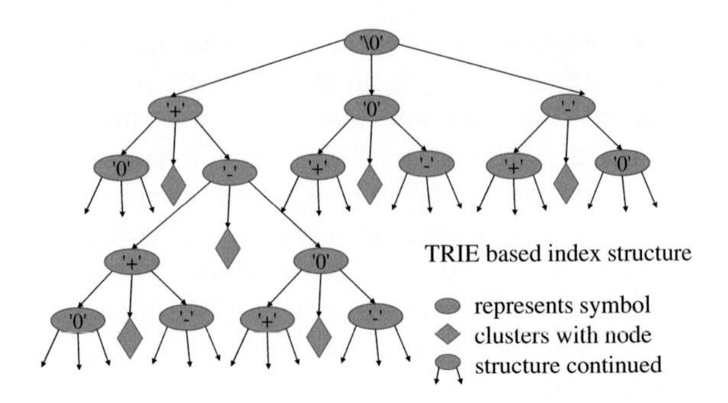

Fig. 2 TRIE-based index structure

Besides storing clustering results obtained above, we also compute and store meta data. This meta data is required to be loaded before execution of any query on trie-based structure as shown in Fig. 2. In trie-based structure, root node of the trie is assigned NULL, and it has three child nodes, one for each logical symbol that can occur at the start of a logical pattern. At every next level of the trie, every node has three child nodes, two of them are used for next symbols of a logical pattern and the third node (in the middle as diamond shape structures in Fig. 2) is used for storing information about the logical pattern that starts at root node and ends at current node.

Diamond node at the end of every logical pattern contains three pieces of information: first about clusters, locations of clusters and normalization factors used during clustering. Cluster information includes centroid, radius and list indexes of runs appeared in it and this is being used in computing heuristic values, i.e., in making a choice which cluster to load next. Location of cluster is stored using two figures, which indicate number of bytes to skip, and number of bytes to load from the corresponding logical pattern file. Normalization factors are used to normalize duration and values of query before comparing it with clusters during search.

4.3 Query Language for Logical Patterns

As stated above, an event can be described by simultaneous occurrence of logical patterns across multiple sensors. In the query for every sensor, a logical pattern is specified along with durations and values for every symbol present in that pattern. Value here corresponds to average slopes for '+' and '−' and average magnitude for '0'. These duration and values can take one of following *type*:

1. Free type: This type is represented by '*' and does not require to specify any value. This indicates that user has no preference on any range retrieved for all values defined in this type.
2. Approximate type: This type is represented by a number e.g. '10'. This indicates that user prefer values closer to this number for all values defined in this type.
3. Range type: This type can be represented in three ways, first way is to have a bound on maximum of the average of symbol e.g. '<10', second way is to have

a bound on minimum of the average of the symbol e.g. '>10' and third way is to have bound on both minimum and maximum e.g. '8–12'. It is important to notice that this type specifies strict criterion on values at symbol level not at particular value with in symbol.

By following above guidelines, one can write query for a logical pattern on a single sensor as: "Engine Speed: 0–0 : $*, 3, * :> 30 : 4 - 8 :< 10$". Here, "$*, 3, *$" represents durations of symbols and implies engine speed should be steady '0' with average magnitude value '> 30' for any duration of time and it should fall '−' with average of '4–8' for approximately '3' seconds, and then become steady '0' with average absolute value '<10 for any duration again. Similarly one can define query for multi-sensor by specifying a valid query for every sensor. In this way, one can exploit our language to describe multi-sensor patterns having complex shapes with very high precision.

4.4 Search Algorithm

In this section, we describe our Algorithm 2 used for searching multi-sensor logical patterns against a query, e.g., "$sensor1 : 0- : *, 3 :> 20 :< -5 \& sensor2 : 0+ : *, 3 :< 10 :> 200$". As mentioned in Sect. 4.2, every logical pattern('$0 -'$ /'$0+'$) for every sensor present in a query(sensor1/sensor2) is stored in a separate file having clusters stored in sequential manner with in the folder of corresponding sensor. Now our task is to find common indexes (i.e., indexes at which '0−' occurs in sensor1 and '0+' occurs in sensor2) among all patterns stored in file corresponding to every sensor. In addition to this, logical patterns must satisfy other specified properties (i.e., durations ($*$, 3) and values ($> 20 :< -5$) for sensor1). This problem can be broken down into following two sub-problems:

- Define a heuristic which can be used to decide which cluster to load among all available choices. Our heuristic should direct search to find indexes of specified patterns efficiently.
- Given one cluster from every sensor present in query, find common indexes with properties specified in query.

Algorithm 2: Search Algorithm

```
  Input  : Multi-sensor logical pattern query and TRIE-based index structure loaded in memory
  Output: Hits of multi-sensors logical patterns
1 Initialization: Cluster container list for every sensor;
2 Sort clusters within trie-structure based on distance described in Eq. 3 for every sensor;
3 Load nearest clusters for every sensor in cluster container list;
4 while TRUE do
5 |   Find common time-stamps for new cluster combinations made due to loaded cluster;
6 |   Check termination criterion, stop if required;
7 |   Compute heuristics value for all choices of clusters to load;
8 |   Load cluster based on heuristic value;
9 end
```

Algorithm assumes that TRIE-based index structure is in memory before firing any multi-sensor logical pattern query. In step:1, algorithm starts with initialising a cluster container list for each sensor involved in the query. This list is used to load clusters from file and hold them for searching indexes. Cluster container list is a three dimensional list, where the first dimension is used to store clusters and the second dimension is used store indexes of runs/time-segment and last dimension for indexes of logical patterns with in that particular time-segment. In Sect. 4.2, we note both indexes of runs/time-segments and time-stamps were sorted before storing them so that process of find intersection faster with cluster lists from rest of the sensors can be speeded up. In step:2, we arrange clusters with in trie-structure itself at the diamond node corresponding to the specified logical pattern for every sensor. We compute distance for every cluster between C and Q as defined in Eq. 3 and keep nearest to farthest cluster as a waiting order to load them in memory for every sensor.

Computing distance between cluster C and query Q is not trivial. As Q might have any type among the list mentioned in Sect. 4.3 for duration and values while C always has its centroid as a vector of approximate type for all attributes. Our definition of distance between cluster C and Q in described: 3, which is a product of approximate distance and range distance. Approximate(range) distance is defined only for approximate(range) type of attributes of Q in Eqs. 1 and 2. Note, we don't take account the contribution of free type of attribute in Q in distance function as they don't have specify anything as mentioned in Sect. 4.3. Approximate distance computes euclidean distance for approximate type attributes and range distance estimates chances of not having any logical pattern with in cluster for the range specified using range attributes in query.

$$AD(Q,C) = \begin{cases} \frac{1}{\sum_{i=1}^{2\times M} A_i} \sum_{i=1}^{2\times M} N_i \times (Q_i - Ci)^2, & \text{if approximate variable(s) in } Q \\ 1, & \text{if no approximate variable in } Q \end{cases} \tag{1}$$

$$RD(Q,C) = \begin{cases} \frac{1}{\sum_{i=1}^{2\times M} R_i} \sum_{i=1}^{2\times M} R_i \times \frac{2\times radius(C) - overlap_i}{2\times radius(C)}, & \text{if range variable(s) in } Q \\ 1, & \text{if no range variable in } Q \\ \infty, & \text{if } C \text{ out of range for variable(s)} \end{cases} \tag{2}$$

$$dist(Q,C) = AD(Q,C) \times RD(Q,C) \tag{3}$$

In Eqs. 1 and 2, A_i/R_i is equal to one if ith variable in Q is of approximate/range type and Q_i/C_i are ith variable of query/cluster otherwise zero. M is number of symbols present in the logical pattern for that sensor. $radius(C)$ represents radius of cluster C and $overlap_i$ is the fraction of cluster that lies with in range for ith range type attribute of Q. For example, if cluster has radius 2 and centre at $(2, 0)$ then, for dimension x-dimension i.e. $overlap_x$ will be $0.25, 0.5, 0$ corresponding to >3, 1–3, <0 and for y-dimension $overlap_y$ will be $0.25, 0.5, 0.75$ corresponding to >1, $(-1) - (1)$, >-1. According to Eq. 3, distance measure is ∞ if a cluster is out of range for any of specified range type attribute of query. One must notice that these clusters will automatically come at last in the waiting list of clusters to load in memory. In case if the nearest cluster for any sensor has distance equal to ∞ than there does not exist any match for that hence search process terminates.

In step:3 of our algorithm, we load nearest clusters into memory for every sensor in corresponding cluster container list. In the next step:5 of our algorithm, we look for indexes at which logical patterns specified in query for every sensor occurring together and also satisfies properties mentioned in query. Firstly, we look for intersection in the first dimension of the cluster which represents index of runs/time-segments. In case we find any match than same process is extended in a similar manner to find intersection in another dimension which represents index of logical patterns with in time-segments. While finding match in second dimension, we consider a match even if initial indexes of logical patterns across different sensors have shift below a threshold. We allow such a mismatch for two reasons, (1) response time of different sensors is might be different, specially for events during which vehicle goes through stressful condition and (2) this difference can appear because of uncertainties in the process of encoding time-series data into series of logical symbols.

In step:6 of our algorithm, we check termination criterion (described below), and if we find it true then algorithm terminates at this step. In case it does not then in the next step:7 of our algorithm, we compute heuristic value for all choices of clusters to load as defined in Eq. 4. First term present in heuristic definition computes probability of finding a match between choice of the cluster to load with the previously loaded clusters from the rest of the sensors. Second term is the support of cluster which implies that both first and second terms estimates the expected number of match for the choice of cluster to load collectively. Third term in heuristic definition makes heuristic value higher for those clusters which stands close to values specified in a query. An additional multiplication by $dist(Q, C_{j0})$ is done to get rid of different range of distance value that might come because of different number of attributes present for different types.

$$HV(C_j) = \frac{\left[\bigcap_{i=1,i\neq j}^{K} Index_i\right][\text{List of Files}(C_j)]}{\text{Num of Files}(C_j)} \times Support(C_j) \times \frac{dist(Q, C_{j0})}{dist(Q, C_j)} \tag{4}$$

In step:8, we load the cluster having highest heuristic value computed in previous choice. After that step:5–8 are repeated unless there is a break at step:6, i.e., termination criterion becomes true. In case of new cluster loaded at step:5 of our algorithm, we consider all combinations of this cluster with all previously loaded clusters from rest of the sensors while finding common indexes of logical patterns. Although rare (specially when initially loaded clusters are small) but there is a little possibility of having all clusters returning a zero heuristic value in the initial stages of search. In that case we choose random cluster to load and put clusters from rest of sensors in the waiting list. We also move to the next clusters in the list of clusters for other sensors as well. In the next iteration, we also consider these waiting clusters as choice along with the clusters that are next in list.

Termination Criterion All clusters which are out of range for one or more attribute will have ∞ distance from query, hence they will come together at the end of the waiting list of clusters to load. Once any such cluster arrives in waiting list, we declare the sensor consumed and do not load clusters from that after that. Once all sensors are consumed, we terminate the search process. In addition to this we also keep a counter on number of retrieved instances and stop when number of retrieved instances is equal or greater than it was queried for, e.g., we require the user to ask for the 'top ..' results.

4.5 Ranking Formula

Given query Q has been made for K sensors and for ith sensor logical patterns having M_i number of symbols which implies total $2 \times M_i$ number of attributes i.e. durations and values. We have defined similarity score between retrieved instance I and query Q in Eq. 5. Where $N_{i,j}/R_{i,j}$ are one only if jth attribute of ith sensor is approximate/range type of attribute or else equal to zero, $Q_{i,j}/I_{i,j}$ represents value of jth attribute ith sensor for query/retrieved instance. $Q^*_{i,j}$ is computed for range attributes which is lower/upper limit if bound is on max/min kind and average of both upper and lower limits in case bound on both. T_p is the total permissible shift that initial indexes of different sensors can have and T_I is the maximum difference in initial indexes of logical patterns in retrieved instance I across sensors.

$$SS(Q, I) = \left(\sum_{i=1}^{K} \sum_{j=1}^{2\times M_i} N_{i,j} \times (Q_{i,j} - I_{i,j})^2 + R_{i,j} \times (Q^*_{i,j} - I_{i,j})^2 \right) \times \left(\frac{T_p - T_I}{T_p} \right) \tag{5}$$

5 Experiments and Results

We evaluate the performance of our proposed method on two real life datasets namely, Hard-stop dataset and Laden dataset. we find our algorithm efficient to be applicable to problems mentioned in Sect. 3. First in Sect. 5.1, we provide guideline to tune parameters and perform pre-processing on any real-life dataset so that it best suits to our algorithm. Later in Sects. 5.2 and 5.3, we have also presented both qualitative and quantitative results for both the mentioned datasets receptively. Machine used to produce results presented have Intel Core i5-3470 processor and 4 GB memory.

5.1 Pre-processing and Parameter Selection

Before applying any processing on the time-series data T, there is one optional smoothing step that can be applied in case data is highly contaminated with noise.

Smoothing also leads to higher compression of time-series into symbol series. Moving average filter best suits to our algorithm, as higher smoothing by moving average leads to slow changes in the slope which results in more compact series of logical symbols. One can choose optimal size of moving filter window by visual analysis of sensor data. Larger value of threshold will lead to higher compression and at the risk of loosing significant sensor information.

Similar trade-off also appears for two more parameters namely deviation filter threshold and total permissible mismatch in starting time of logical patterns across different sensors i.e. T_p. Smaller values of both of these parameters will reduce the amount of noise incorporated at the risk of loosing important patterns once for all. Deviation filter threshold again can be fixed by visual analysis. As T_p is not used in off-line processing, hence one can vary it and choose its optimal value by analysing retrieved results against queries for different set of values.

5.2 *Hard Stop Dataset*

We named our first real life dataset as hard-stop dataset. Measurements and statistics about the dataset are depicted in Table 1. This dataset was recorded for 17.5 h approximately over span of 5 days. With in day vehicle was active for several runs with time gaps in between. Only two sensors namely wheel-based vehicle speed and sudden increase in primary cylinder pressure were selected several sensors instrumented on the vehicle. Domain experts define hard-stop event as "sudden drop in wheel-based vehicle speed and sudden increase in primary cylinder pressure".

One such formulated query for hard-stop could be "Wheel-based Vehicle Speed: 0-:*,3:>20:<-5 & Primary Pressure Master Cylinder:0+:*,3:<10,>200". Results were analysed by domain experts and most of them are like hard-stop. In response to this query, 16 instances were retrieved and according to domain expert 43.75 % (7/16) were hard-stop among all and after digging up the rest of the data, they found

Table 1 Hard-stop dataset: time, memory and other statistics of our algorithm at various stages

Measurements	Hard stop dataset
Number of time-stamps	62976
Number of sensors	2
Number of runs/time-segments	126
Size of dataset	3.5 mb
Size of database summary	2.5 mb
Average TRIE size on disk	220 kb
Average TRIE size in memory	22 kb
Average TRIE loading time	39 ms
Average response time for two sensors	9 ms

Note mb, kb and ms implies megabytes, kilobytes and milliseconds

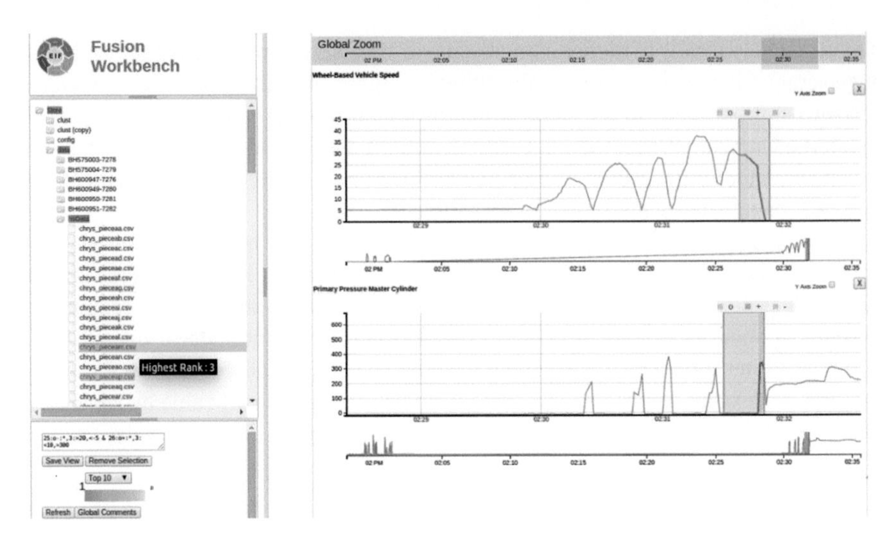

Fig. 3 Retrieved instance of hard-stop

61.11 % (11/18) hard-stops were missed. Before coming to conclusions based upon these numbers, one must acknowledge the brittleness of our algorithm which can change the precision and recall results tremendously. Second important point to notice here is very less response time of our algorithm i.e. in tens of milliseconds as depicted in Table 1.

To allow user to trace retrieved results faster, we have used color coding and highlighting scheme as shown in Fig. 3. On left part of panel in Fig. 3, we highlight both folders and files which contains retrieved instances against the fired query. Every file is given prompt with highest rank that has occurred in that file or folder so that one can locate higher ranked instances faster. In addition to this we also differentiate colour intensity of files based upon highest ranked instance it contains. In Fig. 3, it is clearly visible primary pressure responds earlier(on lower panel) responds for shorter duration in comparison to wheel-based vehicle speed op hard-stop. This behaviour not only reflect capability of our algorithm but also justify assumption of treating sensors separately in queries.

5.3 Laden Dataset

We named our second real life dataset as laden dataset. Measurements and statistics about this dataset are depicted in Table 2. This dataset consists of 481 sessions which was recorded for the duration of 210 h approximately over span of one year. With in these days vehicle was active for several runs with time gaps in between. This

Table 2 Laden dataset: time, memory and other important statistics of our algorithm at various stages

Measurements	Laden dataset
Number of time-stamps	755940
Number of sensors	4
Number of runs/time-segments	481
Size of dataset	27 mb
Size of database summary	85.7 mb
Average TRIE size on disk	212 kb
Average TRIE size in memory	383 kb
Average TRIE loading time	216 ms
Average response time for two sensors	22 ms
Average response time for three sensors	32 ms
Average response time for four sensors	36 ms

Note mb, kb and ms implies megabytes, kilobytes and milliseconds

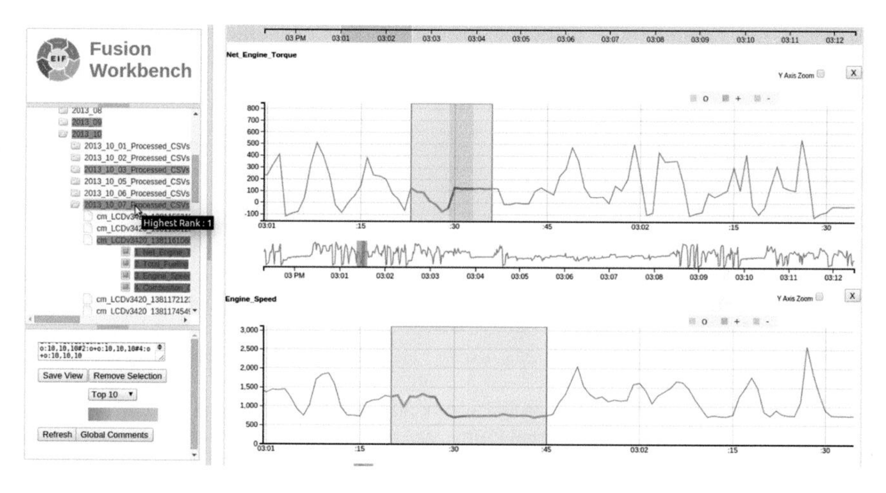

Fig. 4 Retrieved instance of laden condition

dataset similar to hard-stop dataset also comprises of sensor data from engines, and contains instances of a ‘laden condition’ event which is defined by domain experts as high torque and low acceleration occurring together. It is very easy to see and say in Fig. 4, behaviour of torque is similar to engine speed for most of the duration except for highlighted duration. In this dataset, we have selected only four sensors and interestingly response time of our algorithm does not blow up with increase in the number of sensors.

6 Conclusion and Future Work

In this paper, we have proposed a query language that describes multi-sensor events in terms of logical multi-sensor patterns. We have also described a mechanism to encode and store time-series to enable efficient search for such logical patterns. Our approach has multiple use-cases that have been found to be useful in practice: e.g., to build a tool for annotating events and train event classifier on vehicle sensor data or for selecting number of sub-sequences to apply such a classifier during the detection stage, as well for exploratory data analysis. Our algorithm is capable of handling logical multi-sensor patterns which are of variable length and similar shape along with shift in their occurrence across different sensors.

In the future, it would be interesting to investigate the possibility of a temporal rule mining approach e.g. "simultaneous occurrence of some patterns in sensor1 and sensor2 leads to occurrence another logical pattern in sensor3, after time τ". Another interesting area to explore is to use our technique of clustering logically encoded sub-sequences of a time-series as a preliminary step to determine the window-size parameter for various motif discovery or outlier detection algorithms e.g. [15, 16, 27, 28].

References

1. Say, A.C.: Functions representable in pure QSIM. In: Proceeding of Fifth Turkish Symposium on Artificial Intelligence and Neural Networks, Istanbul, Turkey, Citeseer, pp. 251–255 (1996)
2. Evans, P.C., Annunziata, M.: Industrial internet: pushing the boundaries of minds and machines. General Electric, p. 21 (2012)
3. Hassan, E., Shroff, G., Agarwal, P.: Multi-sensor event detection using shape histograms. CoRR arXiv:abs/1408.3733 (2014)
4. Ye, L., Keogh, E.: Time series shapelets: a new primitive for data mining. In: Proceedings of the 15th ACM SIGKDD international Conference on Knowledge Discovery and Data Mining, pp. 947–956. ACM (2009)
5. Hu, B.: Time series classification under more realistic assumptions
6. Fu, T.C.: A review on time series data mining. Eng. Appl. Artif. Intell. **24**(1), 164–181 (2011)
7. Esling, P., Agon, C.: ACM Comput. Surv. (CSUR). Time-series data mining **45**(1), 12 (2012)
8. Ding, H., Trajcevski, G., Scheuermann, P., Wang, X., Keogh, E.: Querying and mining of time series data: Experimental comparison of representations and distance measures. Proc. VLDB Endow. **1**(2), 1081–1094 (2008)
9. Bettaiah, V., Ranganath, H.S.: An analysis of time series representation methods: data mining applications perspective. In: Proceedings of the 2014 ACM Southeast Regional Conference, p. 16. ACM (2014)
10. Lin, J., Keogh, E., Wei, L., Lonardi, S.: Experiencing sax: a novel symbolic representation of time series. Data Min. Knowl. Discov. **15**(2), 107–144 (2007)
11. Moskovitch, R., Shahar, Y.: Classification-driven temporal discretization of multivariate time series. Data Min. Knowl. Discov. **2014**, 1–43 (2014)
12. Krawczak, M., Szkatula, G.: Nominal time series representation for the clustering problem. In: Intelligent Systems (IS), 2012 6th IEEE International Conference, pp. 182–187. IEEE (2012)
13. Rakthanmanon, T., Campana, B., Mueen, A., Batista, G., Westover, B., Zhu, Q., Zakaria, J., Keogh, E.: Searching and mining trillions of time series subsequences under dynamic time

warping. In: Proceedings of the 18th ACM SIGKDD International Conference on Knowledge Discovery and Data Mining. KDD '12 (2012)
14. Gillian, N., Knapp, R.B., OModhrain, S.: Recognition of multivariate temporal musical gestures using n-dimensional dynamic time warping. In: Proceeding of the 11th International Conference on New Interfaces for Musical Expression (2011)
15. Mueen, A.: Time series motif discovery: dimensions and applications. Wiley Interdiscip. Rev.: Data Min. Knowl. Discov. **4**(2), 152–159 (2014)
16. Mueen, A., Keogh, E.J., Zhu, Q., Cash, S., Westover, M.B.: Exact discovery of time series motifs. In: SDM, SIAM, pp. 473–484 (2009)
17. Lam, H.T., Calders, T., Pham, N.: Online discovery of top-k similar motifs in time series data. In: SDM, SIAM, pp. 1004–1015 (2011)
18. Begum, N., Keogh, E.: Rare time series motif discovery from unbounded streams. Proc. VLDB Endow. **8**(2), 149–160 (2014)
19. Agarwal, P., Shroff, G., Saikia, S., Khan, Z.: Efficiently discovering frequent motifs in large-scale sensor data. In: Proceedings of the 2nd ACM iKDD International Conference on Data Sciences CoDS '15 (2015)
20. García, F.T., Villalba, L.J.G., Portela, J.: Intelligent system for time series classification using support vector machines applied to supply-chain. Expert Syst. Appl. **39**(12), 10590–10599 (2012)
21. Lee, Y.H., Wei, C.P., Cheng, T.H., Yang, C.T.: Nearest-neighbor-based approach to time-series classification. Decis. Support Syst. **53**(1), 207–217 (2012)
22. Lines, J., Bagnall, A.: Time series classification with ensembles of elastic distance measures. Data Min. Knowl. Discov. 1–28 (2014)
23. Deng, H., Runger, G., Tuv, E., Vladimir, M.: A time series forest for classification and feature extraction. Inf. Sci. **239**, 142–153 (2013)
24. Batal, I., Fradkin, D., Harrison, J., Moerchen, F., Hauskrecht, M.: Mining recent temporal patterns for event detection in multivariate time series data. In: Proceedings of the 18th ACM SIGKDD International Conference on Knowledge Discovery and Data Mining, pp. 280–288. ACM (2012)
25. Moskovitch, R., Shahar, Y.: Classification of multivariate time series via temporal abstraction and time intervals mining. Knowl. Inf. Syst. 1–40 (2014)
26. Zhang, T., Ramakrishnan, R., Livny, M.: Birch: an efficient data clustering method for very large databases. In: ACM SIGMOD Record, vol. 25, pp. 103–114. ACM (1996)
27. Gupta, M., Gao, J., Aggarwal, C., Han, J.: Outlier detection for temporal data. Synth. Lect. Data Min. Knowl. Discov. **5**(1), 1–129 (2014)
28. Jones, M., Nikovski, D., Imamura, M., Hirata, T.: Anomaly Detection in Real-valued Multidimensional Time Series (2014)

Author Index

D. Ryżko et al. (eds.), *Machine Intelligence and Big Data in Industry*,
Studies in Big Data 19, DOI 10.1007/978-3-319-30315-4